THIS BOOK IS THE PROPERTY OF

THE HABERDASHERS' ASKE'S SCHOOL, ELSTREE

NAME	FORM	REC'D.	RET'D.
G.D. SPENCER	5/35	7/12/71.	23/3/72

THIEVES AND
ANGELS

By the same author

POETRY

Imaginings	Putnam, 1961
Against the Cruel Frost	Putnam, 1963
Object Relations	Methuen, 1967

FICTION

Lights in the Sky Country	Putnam, 1962
Flesh Wounds	Methuen, 1966

CRITICISM

Llareggub Revisited	Bowes and Bowes, 1962
The Quest for Love	Methuen, 1965

EDUCATION

English for the Rejected	Cambridge, 1964
The Secret Places	Methuen, 1964
The Exploring Word	Cambridge, 1967
Children's Writing	Cambridge, 1967

COMPILATIONS

Children's Games	Gordon Fraser, 1957
Iron, Honey, Gold	Cambridge, 1961
People and Diamonds	Cambridge, 1962–6
Thieves and Angels	Cambridge, 1962
Visions of Life	Cambridge, 1964
The Broadstream Books	Cambridge, 1965

Shortened editions of novels for school use:

    Oliver Twist, by Charles Dickens

    Childhood, by Maxim Gorki

    Roughing It, by Mark Twain

    Pudd'nhead Wilson, by Mark Twain

I've Got to Use Words	Cambridge, 1966
The Cambridge Hymnal with Elizabeth Poston	Cambridge, 1967
Mr Weston's Good Wine, by T. F. Powys with an Introduction	Heinemann, 1966
Plucking the Rushes Chinese poems in translation	Heinemann, 1967

THIEVES
AND
ANGELS

Dramatic pieces for use in schools

COMPILED AND EDITED BY
DAVID HOLBROOK
Sometime Fellow of King's College, Cambridge

WITH A FOREWORD BY
BERNARD MILES

CAMBRIDGE
AT THE UNIVERSITY PRESS
1968

Published by the Syndics of the Cambridge University Press
Bentley House, 200 Euston Road, London, N.W. 1
American Branch: 32 East 57th Street, New York, N.Y. 10022

© Cambridge University Press 1962

Standard Book Number: 521 05302 1

First Published 1962
Reprinted 1968

Printed in Great Britain
at the University Printing House, Cambridge
(Brooke Crutchley, University Printer)

FOR THE
MADDERMARKET THEATRE
NORWICH

CONTENTS

vii

CONTENTS

FOREWORD

One evening in the late eighteenth century, when an actor named Griffiths spoke the words in Shakespeare's *Richard III* 'A horse, a horse, my kingdom for a horse!' in a Herefordshire theatre, it is recorded that a local grazier stood up in the stalls and offered to lend him one.

There is also a story told about a performance of *Othello* given by a company touring the mid-west of America in the eighteen-eighties. The audience were simple folk, many of them rough cowboys who had never seen a play in their lives, let alone one so fine. Carried away by the story, and accustomed only to the rough justice of the ranges, in the middle of the play one of the cowboys drew his six-shooter and shot Iago dead. Then, realising what he had done, shot himself dead. It is said that actor and cowboy lie buried in the same grave, beneath a headstone inscribed with the words 'Here lie the perfect actor and the perfect member of an audience'. Stage talk is full of such wonderful stories, many of them quite true.

In 1943 when I was on tour with an Old Vic production of *Othello* we did a week at the Theatre Royal, Hanley. I was Iago. In the third act, where Desdemona asks Iago if he (of all people) can help her to win back Othello's love, the producer arranged for her to be kneeling, with her long hair about her shoulders. One night as, standing behind her and stroking her hair, I spoke the horrible lie 'All may be well', a woman in the audience, her voice full of passion, cried out 'You b....!' Nobody laughed and the play went on.

Later on the tour I invited my landlady to come and see the play one evening. This was her first experience of Shakespeare, and when she brought in my supper after the performance I

ix

asked her what she thought of it. She answered with great seriousness 'Well, it certainly makes your own troubles look very very light'.

It took a long time for actors to be accepted as decent members of society, instead of rogues and vagabonds, but they have always known the strength of their craft, in making the imaginary seem real. Their biggest problem has been to put that craft to the service of worthy material. The dramatic experience goes to the root of living. The cradle of drama is religion, whether in India, China, Greece, or here at home. To *see* the women coming in the first light of Easter and meeting the angels, would strongly confirm the Resurrection story. The angels' question, 'Whom are you looking for here at the sepulchre?' demands the same response as Christian's question, 'What shall I do to be saved?' The drama is a model of life itself with all its comicality and desperation. All we are and all we suffer and enjoy has been thought of before, along with all we never dreamt we could possibly be. We are not alone. There may be no easy solutions but at least we are in company.

Then to *do* it, to act it out, is the greatest of satisfactions; to *be* Mak or one of the shepherds or Mak's wife, to *be* one of the three kings, following the star. To *be* Christian or Hopeful and (even more satisfying) to *be* Apollyon or one of the jury in Vanity Fair, to *be* Falstaff.

Best of all to act out a play of one's own invention, to make it up and then to do it, to take the first steps in understanding and in fixing that understanding with words, these are the first steps of poetry and love. In this book David Holbrook has marked a few of the guide-lines leading to the gateway. Beyond the gateway are meadows and mountains and valleys to be discovered for oneself, in the most exacting and enchanting young company.

BERNARD MILES

THE MERMAID THEATRE
LONDON

x

A NOTE FOR THE TEACHER

The dramatic pieces chosen for this book are intended for various uses in school. The book may be given to children to read, to themselves, or as a source of examples to be imitated in the imaginative composition of drama of their own. In this way the volume may be introduced as a next step following the development of free drama and mime, once the children have reached the stage at which it seems appropriate to begin writing down scenarios and dialogue.

Or the pieces may be read out to children, in the same way that one reads aloud poems, short stories and passages from novels. This kind of work may be followed by oral or written questions, to bring out their comprehension of the meaning. Staging may also be discussed, in relation to the meaning.

The book may also be used for classroom drama, playreadings with action, either in groups each working on the stage interpretation of its own play, or reading the text while a few of the class act the piece. Or, of course, these pieces may be given as school productions. The intention of the book is to link drama with literary work in English, since the six pieces here may be studied as reading matter, and linked with other literary studies (except perhaps for number 5, which is a good story but for which I claim no literary merit).

The 'programme' in Appendix I should suggest a whole scheme of semi-dramatic projects, to link poetry, music, drama and ceremony—simple dance episodes could, for instance, happily be introduced. Appendix II should help link the ceremony of early childhood—if Mr Punch may be called ceremony (perhaps he is anti-ceremony?)—with more literary dramatic work later. This piece is also the classic of puppetry, and has elements of the old Italian spontaneous *commedia dell'arte*: this script should be treated with disrespect and altered to suit local circumstances with contemporary allusion, as with music hall. I put Punch in the appendix because, although he may be unworthy of literary attention, he is a great myth, and no child should grow up without making his acquaintance. For considerations of other links between the intuitive dramatic faculty of childhood and drama, ceremony and the literary consideration of theatre, see the present compiler's *Children's Games* (Gordon Fraser).

In the plate is shown a plan of the special classroom ('The Mummery') at the new Perse School, Cambridge. This simple device of putting a stage in the English room may perhaps more widely recommend itself to those who plan schools—it costs comparatively little, but is a great incentive to the child's dramatic imagination. I discuss at length elsewhere (see *English for Maturity*, pp. 202 ff.) what I think the development of the dramatic faculty in our society could do for our civilisation. If we could replace the old flower-festivals, well-dressing ceremonies, mummings, sainings, May-ceremonies and mystery plays with new forms of dramatic-poetic ceremony fit for modern times we might be brought to possess something of the satisfying sense of community, liberating sympathy, which animated the traditional English village of the past. Of course it would also exert on us the need to have something to celebrate...

I make no elaborate suggestions for production in my notes on the plays. The best productions (I say as a sometime actor of minor parts in my youth at the Maddermarket Theatre, Norwich) are the simplest, the least obtrusive. A great work of dramatic art does not need propping up by stage business and bewildering sets—the best examples to keep in mind are perhaps the pictures by the earlier great masters of such subjects as the Nativity: a simple formal scene as backcloth, sumptuous garments—but the main attention concentrated on the human face and human gesture. With children all should concentrate on the child's possession of the experience in the words, and on making this

A Note for the Teacher

experience a deep, sincere, and sensitively controlled one. The control must come from the work of art, and thus the simpler the mounting of the play is, the less exhibitionist it all is, the less it provides sensational satisfactions for the adults involved, the better, as 'the ceremony of innocence', it will be.

But this true experience does not often come by intention: we can only hope it does come. It came to one of my own children when she, at six, was asked to carry the bowl of bread to the altar at Holy Communion in the village church. Her deep apprehension of the significance of this will never leave her. It is this kind of experience that drama can and should provide, and I hope that here and there by design and accident children in secondary schools may find such experiences through this book.

DAVID HOLBROOK

NOTES ON THE PLAYS

It was suggested to me, as I completed the compilation of this book, that I should preface each play with a note for the teacher. This I have preferred not to do, because I would not wish to impose on a teacher any single interpretation or approach—a teacher must teach in his own way. For any teacher bold enough to use this book will see in these dramatic pieces what use he can make of them, in his whole programme, to suit the stage of development of each group of pupils. He will devise his own strategy and tactics.

But perhaps it may reinforce the confidence of the English teacher if I try briefly to justify my choice of each play for a volume to be used with children in schools.

The Bird-catcher in Hell seemed to me a useful introduction for young children to non-naturalistic drama. As with so much from the Far East (much of which has been so well translated for us by Arthur Waley) the oriental vision touches the vision of children, in its naivety and depth, directly. At one level this is a playful little farce about going to hell, of the kind children inherit in their games about witches and Old Roger. Deeply beneath the surface there is a metaphysical preoccupation with the value of life itself, an irony directed at ethical codes about killing living creatures. And here this little play touches the centre of all ethics—what value do we put on life? The irony, of course, is cast at human hypocrisy about the destruction of life while we continue to relish cooked flesh. This it extends to the gods. All this children would relate in their direct naivety to their own fears of death and their feelings about eating dead flesh, which often seems to them so terrible.

There is thus plenty to engage the child's deepest and unspoken feelings, as dramatic experience. But at the same time this play offers many useful aspects of dramatic art which happily marry with the child's natural intuitive sense. There is, for instance, the opportunity to enter into the character part of the cunning bird-catcher, an elementary Autolycus figure. The whole small play makes a simple dramatic point in such a way that children may discuss it as an exercise in comprehension. It is, too, the kind of simple dramatic episode they could compose in class discussion and develop into a play of their own. The piece also suggests a non-conventional stage—not the proscenium arch kind, nor yet the film- or television-set kind. The Nō stage is formal, with certain parts of the acting space representing higher or lower states of being, and this kind of formal, significant staging is found too in ancient ritual, Greek drama, medieval French religious drama, and English medieval mystery, as with Hellmouth at one side and Heaven at the other. Such symbolic use of space is an important aspect of dramatic training: this may be a simple introduction to it. And the little play suggests the use of space in an arena or in a formal series of constructed levels—open-stage work—at the very beginning. Of course children themselves use space in these ways in their traditional games. With this use of space this little play also suggests formal gesture and formal speaking, as with the Nō choruses.

A teacher with an enthusiasm for literature may like to read in relation to work on this play Yeats's essay on *The Nō Plays of Japan*, and (if he can get hold of a copy) Ezra Pound's and Ernest Fenollosa's *Nō or Accomplishment* (1916). The use of music and mime are also suggested by the form of this play, and these, with the suggestion too of the possible use of masks, may invite co-operation in the school between several departments. Withal, the little play will stand this attention because its point is soundly comic enough, established as it is in metaphysical apprehension.

The Wakefield Second Shepherds' Play (from the Towneley Manuscript) has the same metaphysical depth, because it touches on our deepest preoccupations with birth, and the significance of spiritual rebirth. The sheep play, at its deepest level, contemplates mortality—Mak faces death for his stomach's sake, the sheep is a carnal object which in being translated into a new-born baby nears its own death. All this casts irony on the reality of human birth, mortality and death: but Mak only suffers a comic mock-death, and this relieves our fears of death before we approach the scene of the rebirth of man in the Christ-child. By being brought to proximity with the common corruptness and baser needs of man ('We reap as we sow', says Mak of sexual desire: and the appetite for meat is in the dramatic poem related to this underlying theme), the birth of Christ, the Lamb of God, is given a profound signifi-cance here, simple and moving—this simple mortal-born Embodi-ment of the Creator guarantees the immortality of all men, not least the simplest and commonest. It is a deeply democratic play—emphasising the best elements of communism inherent in Christianity, that all are equal in the sight of God.

All this children apprehend at the unconscious level: and consequently they enjoy this play immensely. It is one of the most remarkable dramatic pieces in the language, and any teacher who works with it will discover the many subtle dramatic points, lending themselves to spatial and carefully timed interpretation. The most obvious example is the Shepherds' return to the hovel, as discussed in *English for Maturity*.

The play has the advantage of being in verse, verse which suggests its own practical value immediately to children—the rhymes help the actor to remember the lines, and the structure guides his emphases. I have tried to render this guiding rhythm in my modernisation—not always with success, I fear. But children easily pick up the vitality of this play, of the goodhearted banter between the shepherds, the assumed bragging of Mak, the domestic brawls, and the deeply stirring simple devotion of the Shepherds' offering speeches. I have been moved to laughter and tears by children naturally entering into the spirit of this medieval master of dramatic poetry.

7

Yet while the banter and gibing is often strangely modern, the movement of the whole is formal and controlled by poetic-dramatic symbolism. The actors must only move and group themselves according to the movement of the verse and its progress. All the sheep-in-cradle business, for instance, must never degenerate into silly farce, but maintain its pace towards the supreme moments, the discovery, the frenzied pretence of Gill that the animal is her baby-changeling. (Children will take the point, as the medieval audience would have done, that if this *is* a child it is a child of the Devil, the 'horned one', and strangely different from the Christ-child whose predicament it echoes.) Careful attention to character and timing will ensure that these points come over.

The set demanded is surely an open one, which, however rich the costumes and enchanting the lighting, remains unrealistic—the hovel, the moor, and the stable at the inn being within the same arena—indeed, this emphasises the point that Christ came down, and his angels appear, within the circumference that includes hardship and common sin. Thus while the sheepskin and crooks need to be carefully 'real' and period, the walls of the hovel should not be solid, and the presence of the flocks need only be suggested by two movable wattle hurdles on the floor of the stage-space.

Great care must be given to the words, and to pointing—of the proverbial expressions ('Many are so wrapped and *are* rogues within'—emphasising the theme of appearance-versus-reality which pervades the play), and of comic lines

> We must drink as we brew...
> That this be the first meal I shall eat today...
> That I eat this child
> That lies in this cradle...
> ...as loud as our sheep smelt...
> When he wakes up he skips...

and so forth. And a seriousness must be preserved, even in the comedy—the business of the sheep in bed is part of that wider 'sex education' which children may get from literature (here to imagine themselves into a situation such as that of a peasant woman in childbirth), which helps them grow up. To make this 'guyed' in any way, in the prurient television variety-show mode, would be to damage an important fantasy experience. At a performance of this play at Bassingbourn I have seen children deeply enjoying this comedy, with all its disturbing sexual irony, in the right true way, without anything 'sexy' or jeering about it: fortunately we had a self-possessed country girl of fifteen for the part of Gill, who spoke with unmitigated robustness of her belly, and who, because she looked after a whole family at week-ends in the absence of her mother, could enter the part of the peasant housewife with real feeling.

Abraham and Isaac is again close to the child's deeper unspoken fears, and possibly the element in this story of a kind of inverted Oedipus myth made it of such great appeal to the medieval mystery dramatists—there are several versions, and this one may be compared with that from Dublin and with the Chester version. The interest is in the 'test', and the balancing of the value of the primal relationship between father and son, with that of both to God. What is considered, really, is the tragic question whether there is anything in human existence or spiritual reality that transcends mortality and mortal bonds—this is what the *Book of Job* asks, in another fashion. In simple psychological terms it allows the child to contemplate the ultimate violence against itself, either from a parent, or from 'the hand of God', within a formal framework of fantasy.

All this a child involved in the play will take in, without its requiring to be made explicit. The drama teacher's sole concern is to make the richest experience of the play possible for the pupils involved. Production, whether in the classroom or on a stage, will obviously require a formalised set—in this instance as near the simple 'pageant' stage as possible, with a place for 'heaven' above, for God and the angel, and a two-level stage for 'home' and 'this hill'. The piece is balletic in its formal movement, and attention is thrown on the words, which develop an almost musical structure, such as T. S. Eliot uses in *Four Quartets*. The mounting tension to the crisis here:

ABRAHAM.   Indeed, sweet son, I cannot tell you, no:
            My heart is now so full of woe.
ISAAC.   Dear father, I pray you hide it not from me,
            But some of your thoughts tell me.
ABRAHAM.   Ah! Isaac! Isaac! I must kill thee!

is like an approaching crisis in music—passion controlled by the formal structure of rhyme and rhythm.

Thus the most important aspect of this small play is that the words must be clearly and carefully spoken, with due attention to the subtleties of rhythm and mode. There is 'character' in the presentation of the son and the father, each being given a nature of his own, but even such distinctions between persons are made formal, by differences in the verse, as if each had a musical theme, as in opera. Everything concentrates on the creation in us of an intolerable passage of suffering and ultimate submission to God's will. Gesture and movement must therefore be at a minimum, to throw all attention on to the moment when Abraham's sword is seized by the angel, expressing the goodness of God, and the security of those that trust in him.

This play then makes a useful exercise in completely non-realistic drama, and in the expression not of personal suffering (though that is there) but of spiritual truths. As an example it could be useful to suggest the composition of plays which enact a simple moral argument.

The Pilgrim's Progress I include with some sinking sense of the remoteness of Bunyan's allegory from the modern sensibility. Perhaps as a drama it may be possible to bring Bunyan to life for our young generation—Bunyan who for centuries was the popular reading-matter of the English people. In a sense it was necessary for Bunyan to go, in order to make room for the new optimism of industrial commercialism, which persuades us to amass possessions in prosperity: the implication of Bunyan's work—

alien to this—is that nothing is worth possessing, because we cannot take it across the river of Death.

The modern world, however, is likely to come to understand Bunyan as the material success of industry and commerce leaves us increasingly without spiritual challenge. Bunyan was expressing, in fact, a profound and universal truth, not so much about adherence to the Baptist (Open Communion) Church, but about the path to personal integration which every man must suffer. Our tragic condition is inescapable, we grow old and die, and our friends, family, and possessions are, ultimately, no comfort to us. We die, as we live, alone, to reverse a phrase of Conrad's. Thus Everyman must, willy-nilly, come to perceive the wrath to come, and flee it: but he flees towards a renewed acceptance of reality, including his mortality, and, eventually, his death and dispossession. The stages of this path to what we would nowadays call 'personal integration' are full of gloomy awarenesses, fiendish assaults and denials, lions by the way, indolences, temptations to believe in the World, Success, and Vanity Fair. The only true path is that which refuses deceptions and self-deceptions, from those of Demas to those of By-Ends, who will seek to arrive by devious means. Ignorance, in the book, does arrive by an 'alternative route'— and is thrust into the Fiery Hill by the angels.

A path to the wicket-gate today may take the form of psychoanalysis, or of the discovery of the self and balance through art, or of a quest in personal relationships or spiritual experience. In D. H. Lawrence it takes the form of exploration of the nature of the human 'mystic now' through love, as may be followed in *Look! We Have Come Through!* But whatever form the quest takes, to seek truth in life, we either follow it with courage or 'rot on the pavement', to use Keats's terms. In fact we may, without spiritual challenge, lose balance and become less able to live. It is thus a vital civilised experience when one is a child—an experience too many of our children grow up today without—to discover that such a thing as spiritual challenge exists. The commercial world (perhaps unconsciously, though not without responsibility for what it is doing) suggests all the time—through advertising, by the creation of optimism, and by deliberate lies about the nature of human life—that there is no quest, no challenge, no tragic exigency on us to take up our pilgrim's staff. Hence much of our spiritual malaise.

Bunyan offers no compromise. He offers only the sentences which 'lay like a millpost at my back'. He offers only the truth of living experience. He represents the disciplines of enquiry and attention to the defining, exploring, metaphorical word which has been the great English heritage since the Reformation—in poetry, liturgy, folk-song and idiom. If, then, we can give our children a little taste of Bunyan we shall offer them a view of another world 'beyond the self', yet more real than the easy sex, frilly petticoat and candyfloss one which lures from the street.

This play, which was most successful many years ago at the Maddermarket Theatre, Norwich, I have cut heavily for school purposes. It includes a large cast, which means that one may involve all the children in a group—in a performance or reading. Or episodes could be used for illustrative purposes in drama or English work, or comprehension work. The piece would make a good school play, particularly if music were used and the set and costumes were well made. Here is an opportunity for 'effects'—as of the mountain emitting fire, Apollyon, the lions, the ascent of Faithful to Heaven, the River Crossing. And it would provide the opportunity for effects which are in the service of a non-realistic drama—they are symbolic effects as in allegorical paintings: as in, say, Breughel.

The Maddermarket Theatre, Norwich, ran for four decades showing mostly such non-conventional drama as this, all Shakespeare (twice), and much poetic drama. The work of this theatre, under the direction of Nugent Monck, belonged, like the Abbey Theatre, Dublin, to the work of the vital standards of European theatre, quite outside the dreary backwaters of English drama in 'rep' and the West End. The Theatre maintained itself, with amateur players working on a play a month: it gave the lie completely to all vindications of rubbish in the amateur and repertory theatre on the grounds of 'giving the public what it wants'. The most regular attenders at the theatre were the owners of the local fish and chip shop, and most people in the auditorium were non-literary, 'non-arty' people. Monck's persistent genius was known abroad, and, in the end, recognised in Norwich. Significantly, when he died, *The Observer* did not record the fact—for Monck's kind of success was that of Bunyan, and not that of Mr Worldly Wiseman.

Death in the Tree I have put into loose, colloquial modern English as an attempt to give the moral neatness of the Shrovetide popular medieval play of Hans Sachs without the tedious regularity of the rhyming couplets as in previous translated versions of this author. Here the pupils can indulge in character a good deal, within the bounds of verse still, without losing the moral point and the controlling fable purpose. The staging, again, should obviously be free and simple. And as an example for creation this is useful. It can be read with success to children in lower streams, too.

Falstaff at Gadshill will, I hope, do two things. First, it should suggest the possible use of fragments of our greatest dramatist which the teacher may himself adapt for secondary school purposes. Secondly, it should draw attention to the magnificent prose of Shakespeare, and its vitality, and provide a touchstone against which to set the flat tedious language of our time. It will be difficult for children brought up on the *Daily Mirror* and *Mirabelle* to get their tongues round these passages from *Henry IV, Part I*, but it will do them no harm, and here 'speech training' can be combined with the need to convey a bold, exciting, comic story. The dramatic interest, once the prose is well and actively spoken, never flags, and surely after taking part in this, or watching it, no child could pronounce Shakespeare dreary? This is an approach to Shakespeare for the non-academic child which I hope may convince many that the inheritance of the Elizabethan artisan is theirs as much as it is that of the academic student.

I need not, I am sure, give laborious attention to the moral implications of the play, and the subtle irony which both makes us sympathetic to Falstaff's lying, bragging and deceit—because he embodies our weaknesses—and at the same time makes us find them untrustworthy and, as Henry does ultimately, deathly, mortal—as mortal as his gout and his pox.

All these fragments of drama have a moral point, and so, even, in his essential irony, has Mr Punch. The collections of fragments I suggest as 'programmes', such as the collection on *Childhood*, should attempt to convey a point, too—an aspect of human experience, without attempting propaganda or heavy moralising. The morality of drama must be the morality of art, and, interestingly enough, many of the episodes collected here set out to be didactic, but are in fact interesting and valuable because of the sympathetic humanity which springs from them incidental to the main didactic purpose. In our time there is much confusion over the question of morality and art: but for our purposes we may turn to the addiction of the child to moral purpose in its play: even in 'Cops and Robbers'. Melanie Klein makes it clear in *Our Adult Society and its Roots in Infancy* that fantasy is an essential part of the growth of the mind in childhood and throughout life, and that in this fantasy, from the very beginnings of identification with the good mother, the child develops moral discrimination. The child needs to learn good from bad, as an exercise in living. And this children do, of course, in their play, which includes, too, a good deal of experiment in cheek and wickedness—as in *Punch*.

The dramatic work I propose here is intended to carry the young mind from its traditional play drama in the playground, and its 'free' imaginative drama in the classroom, into written poetic and allegorical drama where it may exercise itself in the recognition of more mature emotions such as awe, despair, reverence, fear, derision, love, submission, flamboyance, and so forth, within the controlling vision of the dramatic artist. Above all I have tried to make this experience a valid one in the child's own language, and within the compass of his own immature powers.

I have sorted through a great deal of material, and rewritten and revised my texts, in order to make sure that the teacher who takes part in this work may feel he is not wasting his time.

I hope he may feel he is contributing at times deep experiences which persist throughout a child's life, as many of mine at the Maddermarket Theatre have persisted throughout my own years.

own language, and within the compass of his own immature powers.

I have sorted through a great deal of material, and rewritten and revised my texts, in order to make sure that the teacher who takes part in this work may feel he is not wasting his time.

I hope he may feel it is contributing at times deep experiences which persist throughout a child's life, as many of mine at the Haddonmarket Theatre have persisted throughout my own years.

1

The Bird-catcher in Hell

THE CHARACTERS

YAMA, KING OF HELL   KIYOYORI, THE BIRD–CATCHER

DEMONS   CHORUS

YAMA.   Yama the King of Hell comes forth to stand
          At the Meeting of the Ways.
                   (*Shouting*)
          Yai, yai. Where are my minions?

DEMONS. Haa ! Here we are.

YAMA.   If any sinners come along, set upon them and drive
          them off to Hell.

DEMONS. We tremble and obey.

             (*Enter the Bird-catcher*, KIYOYORI)

KIYOYORI. ‘All men are sinners.’ What have I to fear
          More than the rest?
          My name is Kiyoyori the Bird-catcher. I was very
          well known on the Terrestrial Plane. But the span of
          my years came to its appointed close; I was caught in
          the Wind of Impermanence; and here I am, marching
          to the Sunless Land.
          Without a pang
          I leave the world where I was wont to dwell,
          The Temporal World.
          Whither, oh whither have my feet carried me?
          To the Six Ways already I have come.
          Why, here I am already at the meeting of the Six
          Ways of Existence. I think on the whole I’ll go to
          Heaven.

DEMON.   Haha ! That smells like a man. Why, sure enough here’s a sinner coming. We must report him. (*To* YAMA) Please, sir, here's the first sinner arrived already!

YAMA. Then bustle him to Hell at once.

DEMON. I tremble and obey. Listen, you sinner! 'Hell is ever at hand,' which is more than can be said of Heaven. (*Seizing* KIYOYORI) Come on, now, come on! (KIYOYORI *resists*)

Yai, yai!

Let me tell you, you're showing a great deal more spirit than most sinners do. What was your job when you were on the Terrestrial Plane?

KIYOYORI. I was Kiyoyori, the famous bird-catcher.

DEMON. Bird-catcher? That's bad. Taking life from morning to night. That's very serious, you know. I am afraid you will have to go to Hell.

KIYOYORI. Really, I don't consider I'm as bad as all that. I should be very much obliged if you would let me go to Heaven.

DEMON. We must ask King Yama about this. (*To* YAMA) Please, sir—!

YAMA. Well, what is it?

DEMON. It's like this. The sinner says that on the Terrestrial Plane he was a well-known bird-catcher. Now that means taking life all the time; it's a serious matter, and he certainly ought to go to Hell. But when we told him so, he said we'd entirely misjudged him.

What had we better do about it?

YAMA. You'd better send him to me.

DEMON. Very well. (*To* KIYOYORI) Come along, King Yama says he'll see you himself.

KIYOYORI. I'm coming.

DEMON. Here's that sinner you sent for.

YAMA. Listen to me, you sinner. I understand that when you were in the world you spent your whole time snaring birds. You are a very bad man and must go to Hell at once.

KIYOYORI. That's all very well. But the birds I caught were sold to gentlemen to feed their falcons on; so there was really no harm in it.

YAMA. 'Falcon' is another kind of bird, isn't it?

KIYOYORI. Yes, that's right.

YAMA. Well then, I really don't see that there was much harm in it.

KIYOYORI. I see you take my view. It was the falcons who were to blame, not I. That being so, I should be very much obliged if you would allow me to go straight to Heaven.

YAMA. (*reciting in the Nō style*)
Then the great King of Hell—
Because, though on the Hill of Death
Many birds flew, he had not tasted one,
'Come, take your pole,' he cried, and here and now
Give us a demonstration of your art.
Then go in peace.

KIYOYORI. Nothing could be simpler.
I will catch a few birds and present them to you.
Then he took his pole, and crying
'To the hunt, to the hunt!'

CHORUS. 'To the bird-hunt,' he cried,
And suddenly from the steep paths of the southern side
    of the Hill of Death
Many birds came flying.
Then swifter than sight his pole
Darted among them.
'I will roast them,' he cried.
And when they were cooked,
'Please try one,' and he offered them to the King.

YAMA. (*greedily*)
Let me eat it, let me eat it.
(*Eats, smacking his lips*)
Well! I must say they taste uncommonly good!

KIYOYORI. (*to the* DEMONS)
Perhaps you would like to try some?

DEMONS. Oh, thank you! (*They eat greedily and snatch*) I want that bit! No, it's mine! What a flavour!

YAMA. I never tasted anything so nice. You have given us such a treat that I am going to send you back to the world to go on bird-catching for another three years.

KIYOYORI. I am very much obliged to you, I'm sure.

CHORUS. You shall catch many birds,
Pheasant, pigeon, heron and stork.
They shall not elude you, but fall
Fast into the fatal snare.
So he, reprieved, turned back towards the World;
But Yama, loth to see him go, bestowed
A jewelled crown, which Kiyoyori bore
Respectfully to the Terrestrial Plane,
There to begin his second span of life.

THE END

The *Wakefield*
Second Shepherds' Play
of the Nativity

FROM THE

TOWNELEY MANUSCRIPT

MUSIC

Music for this play has been composed by Elizabeth Poston. Enquiries should be addressed to Cambridge University Press.

THE CHARACTERS

1ST SHEPHERD (COLL)	MAK'S WIFE, GILL
2ND SHEPHERD (GIB)	MARY
3RD SHEPHERD (DAW)	THE CHRIST-CHILD
MAK, THE SHEEP STEALER	AN ANGEL

1ST SHEPHERD. *(Enters as if risen from sleep)*
    Lord! What! These weathers are cold, and I am ill-wrapped;
    My hands can't hold, so long have I napped;
    My legs bend and fold, my fingers are chapped.
    Things don't go as they should, and I am all lapped
        In sorrow.
    In storm and tempest,
    Now in the east, now in the west,
    Woe to us who never rest
        Midday nor morrow!

    But we simple shepherds that walk upon the moor,
    Faith, we're nearly homeless and out of the door;
    No wonder, as things stand, if we be poor,
    For the tilth of our land lies as fallow as a floor.
        Why? I'll tell you then:
    We are so lamed,
    So taxed and slammed,
    We are made hand-tamed
        By rich gentlemen.

    Thus they deprive us of rest, Our Lady them harry,
    These men that are Lord-pressed, they cause the plough tarry.
    Some men say this is all for the best: we find it contrary:
    Good husbandmen are opprest to the point of miscarry

In life.
Thus they hold us under,
Thus they bring us to blunder:
It were great wonder
 If we ever should thrive.

If a man gets a lord's livery coat nowadays
Woe to him who answers back what he says:
No one dares oppose the mastery he has
And yet none can believe one word he says,
 No, not a letter.
He can requisition from us
With boast and insolence
And all through his subservience
 To men who are greater.

Along comes any hired man, proud as a crow,
He must borrow my haywain, my plough also,
And I feel the chain—I must give ere he go.
Thus we live in pain, anger and woe,
 By night and day:
He takes as if it belonged—
If I think I'm wronged
I were better hanged
 Than once say him nay.

(He crosses as if to go to his flock)
It does me good, as I walk thus on my own,
Of the world to talk in this manner, and moan:
To my sheep will I walk and listen anon,
There to rest on a baulk or sit on a stone
 Right soon.
For I know, by Our Lady,
If true men they be
We'll get more company
 Ere it be noon.

The Wakefield Second Shepherds' Play of the Nativity

(*He sits as if watching his sheep. Enter* 2ND SHEPHERD *to him*)

2ND SHEPHERD.   *Benedicite* and *Dominus!* what may this mean?
   Why fares this world thus? It's the worst ever seen.
   Lord, these weathers spite us, and the wind is full keen,
   And the frosts so hideous my eyes water with pain,
      No lie.
   Now in dry, now in wet,
   Now in snow, now in sleet;
   When my shoes freeze to my feet
      It is not all easy.

   But as far as I wend, wherever I go,
   I see we married men have much woe,
   We have sorrow, again and again, it falls often so.
   Silly Copple, our hen, both to and fro
      She cackles,
   But begin she to croak,
   To groan or to cluck,
   Woe is he who's the cock
      (*he indicates himself*)
      For he's in the shackles.

   Those men that are wed have not all their will,
   When they are hard rid they sigh full still.
   God knows how they're led full hard and full ill:
   In kitchen or bed they say nothing: they're well
      Tongue tied.
   My part have I found,
   My lesson is learned,
   Woe is he that is bound
      For he has to abide.

   But now lately in our lives—a marvel to me,
   And I think my heart dives to my boots to see
   How destiny drives some—how should it be?—
   Some men will have two wives, and some men three—

Or more!
Some are woe that have any!
One is too many!
Woe is he that has many,
        For he feels it sore!

But young men who're wooing, for God that you
                                                wrought,
Be well ware of wedding, and think in your thought:
'Had I known' is a saying that serves you not—
Afterwards: much mourning has wedding brought
        And grief.
With many a sharp shower
You may catch in an hour
What shall twist you full sore
        For the rest of your life.

For, as I've read the Epistle, I have one that I fear,
As sharp as a thistle, as rough as a briar.
Her brow is all bristle, with sour lenten cheer,
Once she's wet her whistle she can sing loud and clear
        Her Paternoster.
She's as great as a whale,
Has a gallon of gall.
By Christ who died for us all,
        I wish I'd run till I'd lost her!

1ST SHEPHERD.   God save all here. Are you deaf?

                                                There you stand!

2ND SHEPHERD.   The Devil stick in your maw! You're late, man!
See anything of Daw?

1ST SHEPHERD.   Yea, on a ley land,
I heard him blow: he comes here at hand
        Not far.
Stand still.

2ND SHEPHERD. Why?

The Wakefield Second Shepherds' Play of the Nativity

1ST SHEPHERD.   Because he's coming, hope I;
2ND SHEPHERD.   He'll tell us both a lie
    Unless we beware.

             (Enter 3RD SHEPHERD, a Boy)
3RD SHEPHERD.   Christ's cross me speed and Saint Nicholas!
    Of both I have need: it is worse than it ever was.
    We should take no heed of the world, let it pass—
    Always turning to dread, it's brittle as glass
        And slithers.
    This world never fared so
    Full of freaks as it does now,
    One good day, next in woe,
        All writhes and withers.

    Never since Noah's flood have such floods been,
    Winds and rains so bad, and storms so keen.
    Some stammer, some have stood to doubt what they've
                            seen:
    Now God turn all to good, I say what I mean,
        For ponder—
    These floods they drown
    All in field and town,
    They bear all down
        And we wonder.

    We that walk in the night our cattle to keep,
    See many sudden sights while other men sleep.
             (He sees his companions)
    Yet my heart grows light—I see rogues peep!
    Now you monsters—all right, I'll give my sheep
        A turning. *(He makes to go away)*
    But full ill's what I've meant
    As I've talked here. Repent?
    Me? Lightly: I'm bent
        On toasting my toes, not burning:
             (He stands by the fire)

(*To the other two* SHEPHERDS)
Ah, sir, God save you, and master mine! (*To* 1ST SHEPHERD)
What'll I have? Thanks! A drink, then I'll dine.

1ST SHEPHERD.   Christ's curse, you knave, you're a lazy swine.

2ND SHEPHERD.   What makes the boy rave? You can wait—for
a time:
We've ate it.   (*Threatens him*)
Put your head in the gate!
Though the shrew comes late
Yet he's in such a state,
All ready to dine—but he's had it.

3RD SHEPHERD.   Such servants as I that sweats and stinks
Eats our bread full dry, and get no thanks:
We're often wet and weary, while the master has forty
winks,
Yet he brings it out grudgingly, our dinner and drinks.
But they give plenty of thought
Both our dame and our sire
(When we've run in the mire)
To nipping our hire,
And paying us short.

But hear this master: by the food you pack
I shall work after, according to what's in the sack:
I shall only do a little, sir, and often stop for a lark
For your little old suppers never lay heavy on my stomach
Yet in the field.
(2ND SHEPHERD *makes an angry dive at him: he runs*)
Ah, but why should I care?
I can still jump in the air
And they say 'cheap ware
Gives a bad yield.'

1ST SHEPHERD.   You're a fool of a lad then to ride out to court
With a man who keeps his money short.

30

2ND SHEPHERD. Peace, boy! Enough said: no more jangling
Or I shall get mad, by Heaven's King,
With your whine.
Where are our sheep, boy, are they gone?

3RD SHEPHERD. I've not seen 'em, sir, since this morn
I left them in the corn
When the clock struck nine.

Their pasture is good: they cannot go wrong.

1ST SHEPHERD. That's right, then. Good! (*Pause*) These nights
are long
By the Rood. I wish someone would give us a
song.

2ND SHEPHERD. So I thought as I stood. Who'll help us along?

3RD SHEPHERD. I'll start.

1ST SHEPHERD. Let me sing the tenor,

2ND SHEPHERD. And I the treble, so high.

3RD SHEPHERD. Then the counter-tenor for me:
See how far we get.

(*They sing*)
(*Enter* MAK, *the Sheep-stealer, with a cloak over his smock*)

MAK. Now, Lord, in your heaven, that made both moon and
stars,
And more than I could name even: Your will comes to me
sparse,
I am all uneven, and you twist at my horns.
Now would God I were in heaven; for up there's no
new-borns
To wail all night.

1ST SHEPHERD. Who's that pipes so poor?

MAK. Would God you knew how I fare!
A man that walks on the moor
For whom all's not right.

31

2ND SHEPHERD.   Mak! Where have you been? Tell us the
tidings.

3RD SHEPHERD.   He's come has he then? Everyone look to his
things....

MAK.   (*Pretends to be a nobleman, and speaks with a 'southern'—a
fancy—accent*)

Why, what do you mean? I'm a yeoman of the king's,
The self and the same, a lord's man, pulling strings
    And such, I.
Fie on you! Get hence
Out of my presence!
I must have reverence.
    Why, who am I?

1ST SHEPHERD.   Why, this stuff's quaint! Mak, you're in disguise.

2ND SHEPHERD.   Mak, d'you want to be a saint? There's some
need in your eyes.

3RD SHEPHERD.   He can put on paint—but the devil knows who
to hang.

MAK.   I've only to make a complaint, and you'll all get strung,
    By a twist of my tongue.
Now let me tell *you* what to do.

1ST SHEPHERD.   Oh yes, Mak, is that so?
Take your false teeth out do,
    And put in a bung.

2ND SHEPHERD.   Mak, I can see the devil in your eye: I've a
thwack I'll lend you.

3RD SHEPHERD.   Mak, don't you know me? By God, I'll mend
you.

(*They threaten to beat him.* MAK *quickly drops his lordly pretence
and becomes himself. He shakes hands*)

MAK.   God help you all three. I thought I'd seen you.
You're a fair company.

1ST SHEPHERD.   Can you come clean now?

2ND SHEPHERD.   Makes you weep!

1ST SHEPHERD.   If so late a man goes
    What can we suppose?
    You've a smell to my nose
      Of stealing sheep.

MAK.   I am true as steel, everyone knows that:
    But a sickness I feel, that don't abate:
    My belly's unwell, it's out of state.

3RD SHEPHERD.   'Seldom lies the devil dead by the gate.'

MAK.   Therefore
    I'm sore and ill:
    If I'm struck stock still
    I haven't eaten as much as a needle
      This month or more.

1ST SHEPHERD.   How fares your wife? By your hood how does
                          she do?

MAK.   Lies weltering—by the Rood—by the fire, lo.
    And a house full of brood. She drinks well, too.
    Whatever else that's good the woman can do,
      That's true—
    Eats as fast as she can,
    And each year that comes to man
    She gives birth to another one,
      And some years two.

If I were gentle and richer by far
I'd be eaten out of house, and harbour:
Yet she's a foul spouse when you get near her,
There's no one knows a worse anywhere
    Than my troll.
D'you know what I'd offer?
I'd give all in my coffer
Tomorrow next to proffer
    Prayers for her soul.

(*They all yawn in weariness and become suddenly sleepy*)

2ND SHEPHERD. I think no one's been awake longer in this whole shire

Than me.
          Sleep I'd take, if it weren't risking my hire.

3RD SHEPHERD. I feel cold. I feel naked: can't we make us a fire?

1ST SHEPHERD. I'm weary, over-walked and run in the mire:
    You stay up.

2ND SHEPHERD. No, I'll lie down here
    For I can't keep up more.

3RD SHEPHERD. I'm as good man's son as they are,
    Yet here I'll drop,

But Mak, come hither, between us you can lie.
          (*lies down and makes* MAK *join them*)

MAK. So I have to stay with ye? And then you can say
    To yourselves, he did?
              (*They all go to sleep*)
(MAK *sits up*)
From my head to my toe
Manus Tuas commendo
Pontio Pilato
    Christ's cross me speed.

(*He rises, the pastors sleeping, and says*)
Now's the time for a man that lacks something good
To stalk privately, then, into a fold
And get smartly to work again: but not be too bold—
Else he might get more than a bargain, if it were told,
    And come to an end.
              (*He indicates hanging*)
Now's the time for a spell:
For he needs good counsel
Who wants to do well—
    But has nothing to spend.

34

(*He works a spell on them*)
But about you in a circle, as round as the moon,
Till I've done what I will, until that it's noon,
Shall you lie stock still, until I've done,
And I'll say for this good strong words, one
    Or two to the right: (*He mutters a spell*)
Over your heads my hand I lift!
Out go your eyes, forgone is your sight!
I must make it sure and tight
    And yet...
(*Shouts*)   *Ere it be night!!*
    (*The* SHEPHERDS *do not stir*)

Ah ha! There, they sleep hard: you can see that.
I was never a shepherd, but now I'll learn it:
If the flock be scared I'll crouch like a dog, flat.
This beast coming hitherward will go well in our pot,
    Fill our bellies tomorrow.
        (*Sheep appears*)
A fat sheep, I dare say,
A good fleece, I dare lay.
I'm a white sheep, when I may—
    But this one I'll borrow.
        (*He steals the sheep and goes home*)

MAK. (*At his own door*)
    Hey, Gill, are you in? Get us some light.

WIFE. Who makes such a din this time of night?
    I am sat down to spin. What hope, if I might
    Leave off, of earning a penny? I hate high and might!
        This is how she fares,
    A house-wife that's made
    To run about like a maid
    Gets nothing done in her trade,
        Always sent on small chores....

MAK. Good wife, open the latch. Don't you see what I'm
                                                    bringing?

WIFE. Wait till I've undone the catch. Ah! Come in my
                                                    sweeting.

MAK. (*Puts down sheep*) Yea, you didn't care much about
                                          leaving me standing.

WIFE. By your naked neck, this'll get you a hanging.

MAK.   Get away!
I can earn my meat,
For in bad times I can get
More than they that slave and sweat
     All the long day.

It just fell to my lot, Gill, I had such grace.

WIFE. It'll be a bit of a blot to be hanged for the case.

MAK. I've escaped before, just as hard distress.

WIFE. But, 'So long goes the pot to the water', they says,
     'At last
It comes home broken.'

MAK. Ah, I know that old token
But never let it be spoken:
     Come on, help me fast.

I wish it were flayed: I could do with an eat;
Near twelvemonth since I had a good feed of sheep meat.

WIFE. They might come before it's dead and hear the sheep
                                                    bleat!

MAK. Then I'd be taken and tried. I'm in a cold sweat.
     Go bar
Gate and door.

WIFE.   Yes, Mak,
So they can't get in the back.

MAK. If they catch me I'll get, from the whole pack,
     The Devil of a scare. (*Pause.* GILL *has an idea*)

36

WIFE.   A good lark have I spied, seeing you think of none:
   This sheep we can hide, till they be gone,
   In my cradle, like a child. Let me alone
   And I'll lie beside in labour, see, and groan;
      Got it?
MAK.   And I'll say you were brought to bed
   Of a boy child this night indeed!
WIFE.   Now my day is made
      By this thing we've plotted.

   This is a good disguise and a long cast
   And a woman's advice proves the best at last.
   I don't care who spies: you go off again fast.
MAK.   I'll get back ere they rise: else blows a cold blast!
      I'll go sleep.
            (*He goes back to the* SHEPHERDS)
   They still sleep, the company,
   And I'll go stalk quietly
   As if it had never been me
      That carried their sheep.
            (*Lies down and sleeps. The* SHEPHERDS *wake*)

1ST SHEPHERD.   *Resurrex a mortrius*: take hold of my hand.
   Judas carnas dominus, I can hardly stand;
   My foot sleeps, by Jesus, my hungry legs bend:
   I thought we had laid us down somewhere near England.
2ND SHEPHERD.   Ah, how are ye?
   Lord! I've slept well
   As fresh as an eel:
   I'm light-headed, I feel
      Like a leaf on a tree.

3RD SHEPHERD.   *Benedicite!* Where have I been? My head quakes,
   My heart jumps out of my skin with the hammer it makes.
   Who made all the din? Oh, my forehead aches.
   I can't stay here within. Hey, fellows, who wakes?

37

We were four.
D'you see anything of Mak now?

1ST SHEPHERD.   We were up before you.

2ND SHEPHERD.   I could take a vow
We'll find him nowhere.

3RD SHEPHERD.   I thought I saw him wrapped in a wolf's
skin.

1ST SHEPHERD.   Many are so lapped and *are* wolves within.

2ND SHEPHERD.   While we all napped I saw Mak with a gin
And a fat sheep, trapped: but he made no din.

3RD SHEPHERD.   Ah, be still—
Your dream drives you mad:
It's but ghosts in your head.

1ST SHEPHERD.   Now God turn all to good
If it be His will.

2ND SHEPHERD.   Rise, Mak, for shame! You lie too long.

MAK.   Now, Christ's Holy name be with us among.
What's all this? By Saint James, something's wrong.
Am I still the same? Ow, my neck's been wrung
        (*He pretends to be cramped: they help him up*)
Hard enough!
Thanks, friends: since suppertime
As I've laid here in grime
I'm in fear of a dream
That's shook my heart rough.

I thought Gill began to croak and travail bad
And gave birth at first cock to a young lad   (*They pat
                                            him on the back*)
To add to our flock. Why should I be glad?
How many more to stand on one rock? Look how many
                                            we've had!

Ah my head!
A house full of young guts—
The devil knock off their nuts!
Woe to a chap with many brats,
    And little bread.

I must go home, by your leave, to Gill, I think.
I pray you look up my sleeve—I steal nothing:
I'd never cause you grief, mates, nor take anything.
                    (*Exit*)

3RD SHEPHERD.   (*To* MAK *as he goes*)   Go on. I bet *you'd*
            starve. (*To the others*) Do you know what I think?
        This morn
    We'd better count over our store.

1ST SHEPHERD.   I'll go on before.
    Let's meet.

2ND SHEPHERD. Where?

3RD SHEPHERD.   At the Crooked Thorn.
                    (*Exeunt*)

MAK.   (*Back at his own door*)
    Undo this door! Who's there? How long shall I stand here?

WIFE.   Who makes such a stir? Go and walk in the moon and stare.

MAK.   Ah Gill. What cheer! It's me, Mak your husband dear!

WIFE.   The devil and his band, will he then still be here,
        Sir Guile?
    He comes croaking through the slot
    As if someone had him by the throat.
    I can't sit at my work, not
        The littlest while.

MAK.   D'you hear what a lather she makes to invent an excuse—
    She does nothing but lark about and scratch her toes.

WIFE.   (*Who hears him, and goes for him*)
    Why! Who wanders, who wakes? Who comes, who goes?
    Who brews? Who bakes? What makes me so hoarse?

And then,
It's terrible to behold
How I work, in hot, in cold:
You'd find it a miserable household
    Without a woman.   (*He gives in*)

But what game have you made with the shepherds,
                                                    Mak?

MAK.   The last word they said as I turned my back
Was they'd look what they had—they'd count over the
                                                    pack.

They won't be glad when they find what they lack
    That they won't, I promise.
But however the game goes
It was me, they'll suppose,
And come here with a foul noise
    And cry out upon us.

But you must do what we agreed.
WIFE.   Ah, that I will,
I'll swaddle him right in my cradle.
Why a greater deceit than this I could pull
I'll lie down straight. Come tuck me in.
MAK.   I will.
WIFE.   And behind!
That Coll and his fellow
They'll nip us full narrow
MAK.   And I shall shout 'Help!' and cry 'Out harro!'
    If that sheep they find.

WIFE.   Listen hard if they call: they'll be here soon.
Come and make ready all, and sing on your own.
Sing lullaby!   (MAK *refuses and she gets up and beats him*
                                                    again)
            Yes, you shall, and I'll groan
And cry out against the wall on Mary and John,

40

And bawl for pain.
Go on! Sing lullaby fast!   (*She threatens him. He croons*)
Trust me: I'm well cast:
If I ever play you false at last,
    Never trust me again.

(*The* THREE SHEPHERDS *meet at the Crooked Thorn.*)

3RD SHEPHERD.   Ah Coll! Good morn! Why, haven't you slept?
1ST SHEPHERD.   I wish I'd never been born—I'm proved a bad
                                                    shepherd—
    A fat wether's been shorn.
3RD SHEPHERD.   Stolen? God forbid!
2ND SHEPHERD.   Who has done us this scorn? Who can we
                                                    suspect?
1ST SHEPHERD.   Someone we know:
    I have sought with my dogs
    All round Horbury Shrogs
    And of fifteen hogs
        Found I but one ewe.

3RD SHEPHERD.   Now trust me if you will—by St Thomas à
                                                    Becket
    Either Mak or Gill is at the back of this racket.
1ST SHEPHERD.   Go on, man, he lay still—how could he take it?
    Saw him go up the hill, with nothing four-legged
        When he left.
2ND SHEPHERD.   Now strike me if I lie,
    If I should this minute die,
    I'd say it was that magpie
        That did this theft.

3RD SHEPHERD.   We'll go to his homestead on our own feet
    And may I never eat bread till I know the truth of it.
1ST SHEPHERD.   And I'll not drink or go to bed till we and Mak
                                                    meet.
2ND SHEPHERD.   I'll not rest or be fed till we all hail and greet

41

Our brother.
The first to find him shout
And keep him in sight.
I shan't sleep easy tonight
    Nor yet any other.

    (*They approach* MAK's *house where* MAK *and* GILL
        are sighing and groaning)

3RD SHEPHERD.   Will you hear how they hack! Listen how they
                                  croon!

1ST SHEPHERD.   Sounds like Doom's last crack—so clean out of
                                  tune,
Call on him.

2ND SHEPHERD.   Mak! Undo your door! Come on!

MAK.   Who's making that racket as if it were noon?
    What riot!
   Who is that I say?

3RD SHEPHERD.   Good fellows: I wish it were day.

MAK.   As far as you may,   (MAK *comes out*)
    Be good chaps, speak quiet

Over a sick woman's head who's ill at ease.
I'd rather be dead than she should get a disease.

WIFE.   (*Shouts*) Go to some other homestead—I'll all of a queeze,
Every foot that you tread nearly makes me sneeze.
    Oh-ah-achoo!

3RD SHEPHERD.   (*Pretends to be very matey*)
Tell us Mak, if you may
How are you, I say?

MAK.   And are you in town today?
    And how are you?   (*Shakes hands*)

You've been in the mire, and you're still wet:
I'll make you a fire if you'll come in and sit.

I want a nurse to hire: remember that?
You remember my nightmare—well, *this is it,*
   In season:
(*Indicates groaning wife*)
I have kids, you know,
Plenty more than enough:
But we must drink as we brew
   And that's but reason.

Won't you eat before you ride? I think you sweat.
2ND SHEPHERD.   No thanks.
                What'll do us good isn't drink.
1ST SHEPHERD.   (*Significantly*) Nor meat.
MAK.   Surely you've nothing to tell but good?
3RD SHEPHERD.   Yes. Our sheep that we keep in gate
Are stolen as they fed. Our loss is great.
MAK.   Sirs! Drink!   (*He hands round pots*)
Had I been there
Someone'd have paid for it dear.
1ST SHEPHERD.   Some say that perhaps you were—
And that's what we think.

2ND SHEPHERD.   Mak, some men think it was Mak, see.
3RD SHEPHERD.   Either you or your wife: *so do we three.*
MAK.   Now, if you suspect my spouse, Gill, or me
Come and rip open our house, and then you'll see
   If we ever had her.
If I have any sheep got
An ewe, ram or what not,
And if Gill has got
   Up since she was laid here—

As I am both true and loyal to God here I pray
That this be the first meal I shall eat today.   (*Points to*
                                the cradle)

1ST SHEPHERD.   Mak—as I wish to avoid Hell—think it over,
 'He soon learns to steal who can't say nay'.

WIFE.   (*Groans*)
  I swelt!
 Get out thieves from my hearthstones!
 Only come to steal, villains!

MAK.   Don't you hear she groans?
  Your hearts should melt.

      (*They go near the cradle*)

WIFE.   Get away thieves from my lad! Don't go near him there!

MAK.   If you knew how she had fared your hearts would be sore.
 It's wrong, and bad, to act like this, where
 A woman's just farrowed: but I'll say no more.
    (*He adopts a pose of injured innocence*)

WIFE.   (*Louder*)
  Ah–ow! My middle!
 I pray to God so mild.
 That if ever I you beguiled,
 That I eat this child
  That lies in this cradle.

MAK.   Peace, woman, for God's pain, and cry not so;
 You'll spill your brains. Why don't you go?
    (*The* SHEPHERDS *confer*)

2ND SHEPHERD.   I know our sheep is slain. What find you, you
             two?

3RD SHEPHERD.   We're working in vain: we may as well go.
  Confound it, mates,
 I can find no flesh
 Salt nor fresh
 In cupboard nor dish,
  Only two empty plates.

 No cattle but this, tame nor wild, (*Indicates baby*)
 None, as I hope for bliss, as loud as our sheep smelt.

44

WIFE. No, so God give me bliss, and joy of my child.

1ST SHEPHERD. We have aimed amiss. Somehow we're be-
                                         guiled.

2ND SHEPHERD.   So have done!
    Sir, Our Lady give it joy,
    Is your child a boy?

MAK. Any lord might enjoy
    This child for his son.

    When he wakes up he skips: it's a joy to see.

3RD SHEPHERD. In good time be his steps and may he be happy.
    But who were his gossips, his midwives, tell me?

MAK. (*Aside*) The devil sew up their lips!

1ST SHEPHERD. (*Aside to the others*)   Listen hard: here's a lie.

MAK.   Sir, God thank them well
    Parkin, and Gibbon Waller, and our tall neighbour
    John Horne: they were all here during labour
    Till you came, making a great row, didn't stay for
    Meat as well....

2ND SHEPHERD. Mak, friends we will be, for we are all one.

MAK. We? No, I'm holding back, me: for me comfort there's
                                         none.

(*Aside*) Farewell all three. We'll be glad when you've gone.
                    (*The* SHEPHERDS *go*)

3RD SHEPHERD. Fair words there may be, but love there's none,
    This year.
        (*They walk gloomily on. Suddenly* 3RD SHEPHERD *stops
                    in his tracks*)

3RD SHEPHERD. Gave you the child anything?

1ST SHEPHERD. I swear not one farthing!

3RD SHEPHERD. Fast again will I fling
    You wait for me here.
            (*He returns to* MAK'S *door*)

Mak! We forgot the child's gift—come and unbar.

MAK.   *(Aside)*   Ah, now they'll rumble the theft—I thought we
were clear!

3RD SHEPHERD.   The child mustn't be bereft, the little day-star.
Mak, by your leave, let me give the little dear
    But sixpence.

MAK.   No! Get away! He sleeps.

3RD SHEPHERD.   I think, yes, he peeps!

MAK.   When he wakes up he weeps
    I pray you get hence.

3RD SHEPHERD.   Give me leave to kiss him and lift him out.
What the devil is this? He has a long snout!

1ST SHEPHERD.   Something's gone amiss: let's not wait about.

2ND SHEPHERD.   Ill spun weft like this always shows in the suit.
    (They go in. He recognises the sheep)
    Ah, I thought so.
That child's like our sheep.

3RD SHEPHERD.   Now Gib, may I peep?
I thought kind might creep
    Where it might not go.

1ST SHEPHERD.   This was a quaint gaud and a far cast,
A really smart fraud.

3RD SHEPHERD.   Yes, sirs, that was't!
Let's brand this bawd and bind her up fast.
A rotten scold hangs at the last.

2ND SHEPHERD.   So shall you!
Look, d'you see how they swaddle
His four feet in the middle!
I never saw in a cradle
    A hornèd lad till now!

MAK.   Peace, I say. Don't say any more.
I'm the father of that: and that woman him bore.

The Wakefield Second Shepherds' Play of the Nativity

1ST SHEPHERD.　What'll you call it? Mak? Lo, God, Mak's heir!

2ND SHEPHERD.　Let be all that! Now God take good care
>   Of your soul hereafter.

WIFE.　　　(*Keeping up the pretence grabs sheep and dances it on her knee*)
>   As pretty a child is he
>   As sits on a woman's knee
>   A dilly-down baby,
>>   To give a man laughter!

3RD SHEPHERD.　I know him by the ear-mark—that's a good
>>>>>>>>>   token.

MAK.　I tell you sirs, hark—his nose was broken.
>   Since then, says a clerk, the fairies bespoke him.

1ST SHEPHERD.　They keep up their false work—time the spell was
>>>>>>>>>   broken.

>   Get a weapon!

WIFE.　　　(*Desperately*)
>   He was took by an elf
>   I saw it myself—
>   As the clock struck twelve
>   He became misshapen.

2ND SHEPHERD.　You two are very deft;—and both in the same bed.

3RD SHEPHERD.　Since they hold to their theft let's do them both
>>>>>>>>>   dead.

MAK.　(*Kneels*)　If I do wrong again ever chop off my head:
>   In your hands be it left.

1ST SHEPHERD.　　　　　　Sirs, do what I said—
>   For this trespass
>   We'll neither curse him nor fight,
>   Bite nor smite,
>   But have done quick
>>   And toss him in canvas.

>   (*They toss* MAK *in a blanket. The* SHEPHERDS *return to their fold*
>   *with the sheep*)

47

1ST SHEPHERD.   Lord, how I am sore, at a point fit to burst.
I can't walk any more: in faith, I must rest.

2ND SHEPHERD.   As a sheep of seven score he weighed in my fist.
I could sleep anywhere, I think, and sleep fast.

3RD SHEPHERD.   Now I pray
Let's lie down on the ground.

1ST SHEPHERD.   I've these thefts on my mind.

3RD SHEPHERD.   Don't chew it all round and round—
Take it easy I say.

(*They lie down and sleep. An* ANGEL *appears above and sings*
Gloria in Excelsis Deo)

ANGEL.   Gentle shepherds, rise, awakened, for now He is
born
To save you from the fiend that made Adam forlorn:
That devil to bind, this night is He born,
God is made your friend now on this morn,
He orders your quests:
To Bethlehem go your way.
There God lies as a baby,
In a crib in poverty
Along with the beasts.

(*Disappears*)

1ST SHEPHERD.   This was the quaintest dream I ever yet heard.
Some marvel it would seem, to be so scared.

2ND SHEPHERD.   Of God's son of Heaven he spoke from on high,
from the Lord,
And the wood in a lightning stream I thought
He made appear!

3RD SHEPHERD.   He spoke as if the child were
In Bethlehem over there.

1ST SHEPHERD.   There's the token—that star.   (*Points to the
sky*)
Let us seek him there.

The Wakefield Second Shepherds' Play of the Nativity

2ND SHEPHERD.   Say, what was his song? Heard you how he
                                                was cracking
   Three breves to a long?

3RD SHEPHERD.   Yea, Mary, he was hacking—
   Not a crotchet wrong, not a beat out or lacking.

1ST SHEPHERD.   Here, let's sing us among, just as he was clacking.
   I can.

(*They sing the* Gloria)

2ND SHEPHERD.   Let's hear you, then, crooners
   You bark-at-the-mooners.

3RD SHEPHERD.   Hold your tongues, groaners:
   Once again, then.

(*They sing again*)

2ND SHEPHERD.   To Bethlehem he said we should go along.
   I fear we wait here too long.

3RD SHEPHERD.   Come, be merry not sad: of mirth is our song,
   Everlasting reward may be ours ere long
     Without an affray.

1ST SHEPHERD.   Let's get there quickly:
   If we be wet and weary
   To the child and Our Lady
     We must not delay.

(*Tries to sing again but is interrupted by the* 2ND SHEPHERD)

2ND SHEPHERD.   We find by the prophecy—let be your din!—
   Of David and Isaiah, and other writers again,
   They prophesied through the clergy, that in a virgin
   Should He come down and lie to pardon our sin
     And slake it,
   Our kind, from woe
   Isaiah said so,
   Ecce virgo
     Concipiet a child that is naked.

3RD SHEPHERD.   Very glad may we be to live on this day
　　That lovely child to see, that Almighty Babe.
　　Lord, I will be well—for once and for aye,
　　If I kneel on my knee—some word to say
　　　　To that child.
　　But the angel said
　　In a crib was he laid;
　　He was poorly arrayed
　　　　Both meanly and mild.

1ST SHEPHERD.   Patriarchs there have been—and prophets in times
　　　　　　　　　　　　　　　　　　　　　　　gone,
　　They desired to have seen this child that is born,
　　They have gone clean away to their graves before he was
　　　　　　　　　　　　　　　　　　　　　　　born:
　　Yet we shall see him, and heaven's queen, before it is morn
　　　　By this token.
　　When I see him and feel
　　His virtue, I'll know well
　　It is true as steel
　　　　What the prophets have spoken:
　　To poor men such as we are that he would appear,
　　First find, and declare by his messenger.

2ND SHEPHERD.   Go we now: let us fare: the place must be near.
　　I am ready. Prepare. Let us go in together
　　　　To that little one.
　　Lord, as your will be—
　　Though we are simple men all three—
　　Grant us some kind of joy
　　　　To comfort your son.
　　　　　　(They arrive at Bethlehem)

1ST SHEPHERD.   Hail, comely and clean: hail young child.
　　Hail, maker of all thing, born of a maiden so mild,
　　You have driven to his den the devil so wild,
　　That false worker of pain, now is he beguiled!

Look he merry is!
Look, he laughs, my sweeting.
A welcome meeting:
I hold here my greeting:
Have a bob of cherries.

2ND SHEPHERD.   Hail sovereign saviour, for us you have sought.
Hail noble child and flower, that all things wrought,
Hail fountain of favour, that made all of nought,
Hail! I kneel and I cower. A bird have I brought
    To my dear.
Hail, little tiny mop,
Of our creed you are crop!
I would drink in your cup,
    Little day-star.

3RD SHEPHERD.   Hail darling dear, full of god-head,
I pray you be near whenever I have need.
Hail! Sweet is your cheer: my heart would bleed
To see you sit here in such poor need
    With no pennies.
Hail! Put out your hand small.
I bring thee but a ball
Have and play thee withal,
    And go to the tennis.

MARY.   The Father of Heaven, God omnipotent
That the world has given, his son has sent:
My name could he name even, and laughed as if he knew
                                his father's intent:
I conceived him through God's might, even as He meant
    And now is he born.
May he keep you from woe:
I shall pray him so,
Tell the world as you go
    And remember this morn.

1ST SHEPHERD.   Farewell lady, so fair to behold
    With your child on your knee.

2ND SHEPHERD.   But he lies there cold:
    Lord, well is me! Now we go forth: behold!

3RD SHEPHERD.   Indeed, already it seems to be told
    To many a crowd.

1ST SHEPHERD.   What grace we have found!

2ND SHEPHERD.   Come forth: now we are sound.

3RD SHEPHERD.   To sing we're bound
    So let's sing loud.

SHEPHERDS.   *(exeunt singing)*
                Explicit pagina Pastorum

THE END

3

Abraham and Isaac

THE CHARACTERS

ABRAHAM

ISAAC   GOD

ANGEL   DOCTOR

ABRAHAM.   Father of Heaven Omnipotent,
      With all my heart to Thee I call;
Thou hast given me land and rent,
My livelihood Thou hast me sent;
I thank Thee highly, evermore, for all.

First of the earth Thou madest Adam,
      And Eve also to be his wife;
All other creatures from those two came;
And now Thou hast granted me, Abraham,
      Here in this land to lead my life.

In my old age Thou hast granted me this
      That this young child with me shall run;
I love no thing so much as this,
Except Thyself, dear Father of bliss,
      As Isaac here, my own sweet son.

I have other children too:
      I love them, but not half as well;
This fair sweet child he cheers me so
In every place where I may go,
      That no discomfort here I feel.

And therefore, Father of Heaven, I pray
      For his health and also for his grace;

55   3-2

Now, Lord, keep him both night and day
That no discomfort nor dismay
  Come to my child in no place.

Now come on, Isaac, my own sweet child;
  Let us go home and take our rest.

ISAAC. Abraham, my own father so mild,
    To follow you I am full glad,
      Both early and late.

ABRAHAM. Come on, sweet child, I love you best
    Of all the children that ever I begat.

(GOD *appears in Heaven with the* ANGEL)

GOD. My angel, fast make thee thy way,
    And on to Middle Earth anon thou go;
  Abraham's heart now will I assay,
    Whether that he be steadfast or no.

Say I command that he should take
  Isaac his young son that he loves so well,
And of his body sacrifice should make
  If any of my friendship he'll keep still.

Show him the way up to the hill
  Where his sacrifice shall be;
I shall assay now his good will
  Whether he loveth best his child or me.
All men shall take example from him
  How they shall keep my commandments.

(*The* ANGEL *descends to Earth*)

ABRAHAM. Now Father of Heaven that formed every thing
    My prayers I make to Thee in dread
  For this day some tender offering
    Here I must make to Thee, O my Godhead.
  Ah, Lord God, Almighty King,
    What act of mine will best make Thee most glad?

If I knew of anything
   You desired, Lord, it would be sped
      With all my might.
To do Thy pleasure on a hill
Truly this is all my will,
   Dear Father, God in Trinity.

ANGEL.   Abraham, Abraham, send you rest.
     Our Lord commandeth you to take
Isaac, your young son that you love best,
   And of his body sacrifice to make.
Into the Land of Vision must thou go
   And offer up thy child unto your Lord.
I shall lead thee, and show also
   Unto God's wish, Abraham, accord,

So follow me upon this green.
ABRAHAM.   Welcome to me my Lord's command,
    And His wish I will not withstand:
    Yet Isaac, my young son in hand,
A most dear child to me has been.

I had rather, if God desired,
   To give up all the goods I have,
Only that Isaac my son should be spared,
   So God in Heaven my soul may save.

I loved never anything so much on earth—
   And now I must the child go kill!
Ah, Lord, my conscience is strongly stirred:
And yet, dear Lord, I am afraid
   To grudge you anything against your will.

I love my child as my life,
   But yet I love my God much more:
For this my heart will waken into strife,
Yet will I not spare child nor wife,
   But do according to my Lord's desire.

Though I love my son so well,
　　Yet smite his head off readily I shall:
O Father of Heaven, to Thee I kneel—
A hard death my son shall feel,
　　To honour Thee, Lord, withal.

ANGEL.　　Abraham! Abraham! This is well said,
　　　　And all these commandments see that you keep:
　　　　But in your heart be not dismayed.

ABRAHAM.　　Nay, nay, indeed, I hold myself well pleased
　　　　To please my God with the best I have.

For though my heart will be heavy, much,
　　To see the blood of my own dear son,
Yet for all this I will not grudge,
But Isaac, my son, I will go fetch
　　And come as fast as ever we can.

Now, Isaac, my own son dear,
　　Where are you, child? Speak to me.

ISAAC.　　My father, sweet father, I am here,
　　　　Making my prayers to the Trinity.

ABRAHAM.　　Rise up, my child, and fast come hither,
　　　　My gentle boy, who is so wise,
For we two, child, must go together,
　　And unto God must make our sacrifice.

ISAAC.　　I am here ready, my father, so.
　　　　Even at your hand I stand right here,
And whatsoever you bid me do,
　　It shall be done with a glad cheer,
　　　　All well and fine.

ABRAHAM.　　Ah! Isaac my own son so dear,
　　　　God's blessing I give you, and mine.

Carry this faggot on your back,
　　And I myself fire shall bring.

Abraham and Isaac

ISAAC.   Father, all these things I will pack,
            I am so glad to do your bidding.

ABRAHAM.   Ah! Lord of Heaven, my hands I wring—
            These childish words all wound my heart!

            Now Isaac, let us go our way
            On to the mountain, with might and main.

ISAAC.   Go, my dear father, and I will try
            To follow you as fast as I can
            Although I am slender.

ABRAHAM.   Ah! Lord, my heart breaks in twain—
            These childish words they are so tender.
                              (*They reach the place*)

            Ah! Isaac, son, now lay it down,
            Drop it from your back, the wood,
            For I must make ready soon
            To honour my Lord God as I should.

ISAAC.   Look, my dear father, there it is.
            To cheer you I kept by you to this place,
            But, father, I marvel much at this—
            Why have you such a heavy face?

            And also, father, all the time dread I
            —Where is your live beast for the kill?
            Both fire and wood we have ready,
            But quick beast have we none upon this hill.

            A live beast, I know well, must be killed
            To make your sacrifice.

ABRAHAM.   Do not dread, I beg you, child,
            Our Lord will send me one for this,

            Some kind of beast for me to take,
            By His sweet angel's hand.

ISAAC.   Yes, father, but my heart begins to quake
            To see that sharp sword in your hand.

59

Why do you hold your sword drawn so?
The way you look much makes me wonder.

ABRAHAM. Ah, Father of Heaven, I am so much in woe!
This child here breaks my heart asunder!

ISAAC. Tell me, dear father, for your ease,
Is your sword bare and drawn for me?

ABRAHAM. Ah, Isaac, sweet son, peace! Peace!
For indeed you break my heart in three.

ISAAC. Now truly, father, why somewhat in your soul
Do you mourn thus more and more?

ABRAHAM. Ah! Lord of Heaven, thy grace let fall,
For my heart was never half so sore.

ISAAC. I pray you father that you let me know
Whether I am to come to harm or no?

ABRAHAM. Indeed, sweet son, I cannot tell you, no:
My heart is now so full of woe.

ISAAC. Dear father, I pray you hide it not from me,
But some of your thoughts tell me.

ABRAHAM. Ah, Isaac! Isaac! I must kill thee!

ISAAC. Kill me, father? Alas, what have I done?

If I have trespassed against you aught
With a stick you may make me mild:
But with your sharp sword kill me not,
For, indeed, father, I am but a child.

ABRAHAM. I am so sorry, son, thy blood to spill,
But truly, child, I may not choose.

ISAAC. Now I wish to God my mother were on this hill:
She would kneel for me on both her knees
To save my life.

And since my mother is not here,
I pray you, father, alter your desire,
And kill me not with your knife.

ABRAHAM.   Forsooth, son, unless I thee kill
I should grieve God right sore, I dread;
It is his commandment and also his will
That I should do this same deed.

He commanded me, son, for certain,
To make my sacrifice with thy blood.
ISAAC.   And is it God's will, then, that I should be slain?
ABRAHAM.   Yea, truly, Isaac my son so good,
And here because of it my hands I wring.

ISAAC.   Now, father, against my Lord's will
I will never grudge, loud nor still;
He might have sent me a better destiny
If it had been His pleasure.

ABRAHAM.   Forsooth, son, unless I do this deed
Grievously displeased our Lord will be.
ISAAC.   Nay, my dear father, God forbid,
That ever you should grieve Him over me.

You have other children, one or two,
Whom you will always love, being your kind:
I pray you, father, make no woe,
For once I am dead and from you go
I shall be gone from your mind.

Therefore do our Lord's bidding
And when I am dead, then pray for me;
But, good father, tell my mother nothing,
Say that I am in another country dwelling.
ABRAHAM.   Ah, Isaac, Isaac, blessed must thou be!

My heart begins strongly for to rise,
  To see the blood of thy blessed body.

ISAAC. Father, since it may be no other wise,
  Let it pass as shall I.

But father, ere I go unto my death,
  I pray you bless me with your hand.

ABRAHAM. Now, Isaac, with all my breath,
  My blessing I give you upon this land,
And God's also I add to this.
  Isaac, Isaac, son, now up you stand,
Your fair sweet mouth that I may kiss.

ISAAC. Now farewell, my own father so fine,
  And greet well my mother on earth.
But I pray you, father, to hide my eyes,
  So that I do not see the stroke of your sharp sword,
That shall my flesh defile.

ABRAHAM. Son, thy words make me to weep full sore;
  Now, my dear son Isaac, speak no more.

ISAAC. Ah! My own dear father, why not more?
  We'll speak together here a little while.

And since that I needs must be dead,
  Yet, my dear father, to you I pray,
Smite but few strokes at my head,
  And make an end as soon as ye may,
And tarry not too long.

ABRAHAM. Thy meek words, child, make me afraid,
  So 'Wellaway!' may be my song,

Only accepting God's will.
  Ah, Isaac, my own sweet child!
Yet kiss me again upon this hill!
  In all this world is none so mild.

ISAAC.   Now, truly, father, all this tarrying
　　　　　It doth my heart but harm;
　　　　　I pray you, father, make an ending.

ABRAHAM.   Come up, sweet son, unto my arm.
　　　　　　I must bind thy hands, too,
　　　　　　Although thou be never so mild.

ISAAC.   Ah, mercy, father! Why should ye do so?

ABRAHAM.   So that you cannot hinder me, my child.

ISAAC.   Nay, indeed, father, I'll not hinder you;
　　　　　　Do to me what you will.
　　　　　　The purpose you have had set you,
　　　　　For God's sake keep it before you still.

　　　　　I am sorry this day to die,
　　　　　　But yet I do not wish my God to grieve;
　　　　　Do as you will to me firmly,
　　　　　　My fair sweet father, for I give you leave.

　　　　　But father I pray you evermore
　　　　　　Tell my mother not this ill;
　　　　　If she knew it she would weep full sore,
　　　　　　For indeed father she loves me well;
　　　　　God's blessing may she have!

　　　　　Now farewell my mother so sweet,
　　　　　　We two seem like no more to meet.

ABRAHAM.   Ah! Isaac! Isaac! Son, you make me to weep,
　　　　　　And with your words you have distempered me.

ISAAC.   In truth, sweet father, I am sorry to grieve you,
　　　　　　I cry you mercy of all that I have done,
　　　　　And of all trespasses that ever I did to you.
　　　　　　Now, dear father, forgive me what I have done.
　　　　　God of Heaven be with me.

ABRAHAM.   Ah, dear child, leave off thy moans,
　　　　　　In all thy life thou grieved me never once;

Abraham and Isaac

Now blessed be thou, body and bones,
That ever you were bred and born!
Thou hast been to me, child, full good:
But, indeed, I mourn thy waste.
Yet here I must needs at the last
In this place shed all thy blood.

Therefore my dear son, here shall you lie,
Unto my task I must go in dread.
I wish indeed that I myself could die
If God would be pleased with such a deed.
And take my body I would gladly offer.

ISAAC. Ah! Mercy, father, mourn ye no more,
Your weeping makes my heart sore,
As much as the death that I must suffer.

Your kerchief, father, about my eyes, please wind.
ABRAHAM. So I shall, my sweetest child on earth.
ISAAC. And please, good father, have this yet in mind
Smite me not often with your sharp sword,
But quickly, once, that it be sped.

(*Here* ABRAHAM *ties a cloth over* ISAAC'S *face*)
ABRAHAM. Now farewell, my child, so full of grace.
ISAAC. Ah, Father! Father! Turn downward my face,
For of your sharp sword I am now in dread.

ABRAHAM. To do this deed I am full sorry,
But, Lord, Thy will I'll not withstand.
ISAAC. Ah! Father of Heaven to Thee I cry,
Lord, receive me into thy hand.

ABRAHAM. Lo! Now the time has come to me
That my sword in his neck shall bite.
Ah, Lord, my heart rises against thee,
I cannot find it in my heart to smite!

My heart seeks not this thing to do,
   Yet fain I would work my Lord's will:
But this young innocent lieth so still,
   I cannot find it in my heart this child to kill.
O! Father of Heaven! What shall I do?

ISAAC.   Ah! Mercy, father, why tarry you so,
   And let me lie thus long on this heath?
Now I would to God the stroke were done!
Father I pray you heartily, make short of my woe,
   And let me not go on imagining my death.

ABRAHAM.   Now heart, why will you not break in three?
   Yet shall you not make me disobey my God.
I will no longer hold my hand for thee,
Because my God would sadly grieved be.
   Now take your fatal stroke, my own dear child.

(*Here* ABRAHAM *raises his sword to strike, but the* ANGEL *appears in the upper part of the stage and seizes the point in his hand and stays it*)

ANGEL.   I am an angel, appearing here betimes:
   From God in heaven I am sent:
Our Lord thanks thee a hundred times
   For keeping his commandment.

He knoweth thy will and also thy heart,
   That thou dreadest Him above all thing:
To make thy misery depart
   A fair ram yonder have I come to bring.

He standeth tied, there, lo, among the briars.
   Now, Abraham, amend thy mood,
For Isaac thy young son that here is
   This day shall not shed his blood.

Go make thy sacrifice with yonder ram!
Now farewell, blessed Abraham,
  For unto Heaven I now go home.
  The way is clear again.
  Take up thy son so free.
    (ANGEL *disappears*)

ABRAHAM.   Ah! Lord, I thank Thee of Thy great grace!
  Now I am eased in many ways!
  Arise up, Isaac, my dear son, arise!
  Stand up, sweet child, and come to me!

ISAAC.   Ah! Mercy, Father! Why smite you not?
  Ah, smite on, father, once with your knife!
ABRAHAM.   Peace, my sweet sir, and take no thought,
  For our Lord of Heaven hath granted you your life,
    By His angel now,
  That thou shouldst not die this day, my son, truly.
ISAAC.   Ah Father, full glad then am I
    Certain, father, I say, certain,
  If this were true.
ABRAHAM.   An hundred times, my son, so fair of face
  For joy thy mouth now will I kiss.

ISAAC.   Ah, my dear father Abraham,
  Will not God be angry that we do thus?

ABRAHAM.   No! No! Never, my sweet son,
  For yonder ram he hath us sent
    Hither down to us.

Yon beast shall die here in your stead
For worship of Our Lord alone:
Go fetch him hither, child, indeed.
ISAAC.   Father, I will go take him by the head
  And bring the beast with me anon.

Ah sheep! Sheep! Blessed must you be
That ever you were sent down hither:
   This day you shall die for me,
In worship of the Holy Trinity.
   Now come fast and go we together,

To my Father of Heaven:
   Though thou be ever so gentle and good,
   Yet I had rather you should shed your blood,
Certain, sheep, than I.

Lo, look here, Father, I have brought
   This gentle sheep, and him to you I give;
But, Lord God, I thank with all my heart
   For I am glad that I shall live,

And kiss again my dear mother.

ABRAHAM.   Now be right merry, my sweet child,
   For this live beast that is so meek
That I shall offer to God before all other.

ISAAC.   And I will fast begin to blow:
   This fire shall burn a full good speed.
   But father, if I stoop down low
   You will not kill me with your sword, will you?

ABRAHAM.   No, truly, sweet son, have no dread—
   That mournful thing is past.

ISAAC.   Yea, but I wish your sword were in its sheath,
   For indeed, father, it makes me still aghast.

(*Here* ABRAHAM *kills the sheep and makes an offering,
kneeling and saying thus:*)

ABRAHAM.   Now Lord God of Heaven in Trinity
   Almighty God omnipotent
   My offering I make to worship Thee
   And this live beast I Thee present.
   Lord receive Thou my intent
   As Thou art God and ground of our Grace.

Abraham and Isaac

(GOD *appears above*)

GOD.   Abraham, Abraham, well may thou speed,
And Isaac, thy young son thee by!
Truly Abraham, for this deed
I shall multiply both your seed
As thick as stars are in the sky.
Both more and less;
As thick as gravel in the sea;
So thickly multiplied your seed shall be,
This grant I for your goodness.

Of you shall come forth many a great one,
And ever be in bliss without end,
For you dread me as God alone,
And keep my commandments every one.
My blessing I give you: world without end.

(The curtains close in heaven)

ABRAHAM.   Lo! Isaac my son, what think you
Of this work that we have wrought?
Full glad and blithe we may be
Against the will of God we grudged not
Upon this fair heath.

ISAAC.   Ah! Father, I thank our Lord God well
That my wit served me so well
To dread God more than my own death.

ABRAHAM.   Why! Dear worthy son, when were you so in dread?
Come little child, tell me more.

ISAAC.   Yea, by my faith, father, now I know indeed,
I was never so afraid before
As I have been upon this hill.
And by thy faith, father, I swear,
I never will again come here
Unless it be against my will.

68

Abraham and Isaac

ABRAHAM. Let it be so: come on, my own sweet son,
And homeward fast now let us go.
ISAAC. By my faith, father, I am ready now,
I was never so willing to go home,
And to speak with my dear mother.
ABRAHAM. Ah! Lord of Heaven I thank Thee
For now I may take home with me
Isaac my young son, alive and free,
The gentlest child above all other,
This I am glad to say.

Now go we forth, my blessed son.
ISAAC. Yes, let us, father, and let us be gone,
For by my troth, had I known
I would never have come away this time.
I pray God give us grace evermore,
And all that we may be worthy for.
                (*Exeunt* ABRAHAM *and* ISAAC)
                (DOCTOR *comes forward*)
DOCTOR. Lo, men and women, now we have shown,
This solemn story to great and small.
It has a point for both scholar and common man,
Even for the wisest of us all,
Without exaggeration.
For this short story you saw here
Shows how we should, with all our power
Keep God's commandments, without deliberation.

I wonder, sirs, if God sent an angel
Commanding you to smite off your child's head
By your good name, is there any of you
That would either grudge or strive against it?

What do you think about it, sirs, today?
Perhaps there are three or four or so?

Women, we know, weep sorrowfully
When that their children die, and go
    As Nature brings their end:
But it is folly, I must here avow
To grudge against God's will, or to grieve you
For he is never to be abused, as we well know,
    By land or water: have this well in mind.

And strive not against our Lord God
    In wealth nor woe, whatever He send
However hard you are beset,
    For when He will, he will amend.

His commandments if you keep with your good heart
    As this play to you has shown,
And faithfully serve Him while you are safe and sound
    That ye may please God night and morn.
Now Jesus who wears the crown of thorn
    Bring us all to Heaven's bliss!
        FINIS.

4

The Pilgrim's Progress

NOTE

The scenes run into each other. They are as follows:

I. *Scene* 1 In the Country
2 Slough of Despond
3 The Wicket-Gate
4 The Cross
5 The House Beautiful

II. *Scene* 6 Apollyon
7 On the Road
8 Vanity Fair

III. *Scene* 9 The Three Gallants
10 Giant Despair
11 The River of Death
12 The Golden Gate

The music required is as follows:

I. 1 Prologue
2 Country Dance
3 Help's Song
4 Angels' Theme
5 Shepherds' Song
(see the beautiful setting in R. Vaughan Williams' *Pilgrim's Progress*)
6 House Beautiful

II. 7 Apollyon's Fight
8 Vanity Fair. Country Dance
9 Pavane
10 Tableau Music

III. 11 Demas Music
12 Despair Music
13 Who would true valour see
14 The River Crossing
15 Holy, Holy, Holy

72

THE CHARACTERS

in order of their appearance

BUNYAN
CHILDREN
LOITERERS
CHRISTIAN
EVANGELIST
CHRISTIAN'S WIFE
HIS CHILDREN
OBSTINATE ⎱ *His neighbours*
PLIABLE ⎰
HELP
MR WORLDLY WISEMAN
GOODWILL
THREE SHINING ANGELS
SHEPHERD BOY
TIMOROUS
MISTRUST
WATCHFUL
DISCRETION
PRUDENCE
PIETY
CHARITY
APOLLYON
AN ANGEL
TWO MEN
FAITHFUL
TALKATIVE
FOUR PEDLARS

COUNTRY-FOLK (*men and women*)
A CROWD OF PEOPLE
NUMEROUS DEVILS (*as merchants*)
LORD LUXURY
LADY LECHERY
SIR HAVING GREEDY
MISTRESS HEADY
LORD CARNAL DELIGHT
LADY VAIN GLORY
MASTER LOVE LUST
MISTRESS MALICE
LADY FEIGNING
MRS NO-GOOD
MR PICKTHANK
MR SUPERSTITION
MR ENVY
HOPEFUL
LORD HATEGOOD
WATCH
MR BLINDMAN (*foreman of the Jury*)
MR NO-GOOD (*No.* 2)
MR MALICE (*No.* 3)
MASTER LOVE LUST (*No.* 4)
MR LIVELOOSE (*No.* 5)

73

MR HEADY (*No.* 6)	MR MONEY-LOVE
MR HIGHMIND (*No.* 7)	MR SAVE-ALL
MR ENMITY (*No.* 8)	DEMAS
MR LIAR (*No.* 9)	GIANT DESPAIR
MR CRUELTY (*No.* 10)	DIFFIDENCE, HIS WIFE
MR HATELIGHT (*No.* 11)	THREE SHEPHERDS AND A BOY
MR IMPLACABLE (*No.* 12)	THE GARDENER
CLERK OF THE COURT	TRUMPETERS
JUDGE	CHORUS OF ANGELS
BYENDS	SHINING ONES
MR HOLD-THE-WORLD	

BUNYAN *in prison as Prologue*

> When at the first I took my pen in hand
> Thus for to write, I did not understand
> That I at all should make a little book
> In such a mode.
>   I, writing of the way
> And race of saints, in this our gospel day,
> Fell suddenly into an allegory
> About their journey, and the way to glory.
> Oh, then come hither,
> And lay my book, thy head and heart together.

ACT I

(BUNYAN *moves to seat by stage, where he acts as prompter*)

BUNYAN. As I walked through the wilderness of this world, I
lighted on a certain place where was a den, and I laid
me down in that place to sleep; and, as I slept, I
dreamed a dream. I dreamed, and behold I saw a man
clothed with rags, standing in a certain place, with his
face from his own house, a book in his hand, and a
great burden upon his back. I looked, and saw him
open the book and read therein; and, as he read, he

wept, and trembled; and not being able longer to contain, he brake out with a lamentable cry, saying, 'What shall I do?' (*He sits*)

(*The middle curtains open showing the field near the City of Destruction. Children are dancing and loiterers watching.* CHRISTIAN *sits apart, R.F.*)

CHRISTIAN. What shall I do to be saved?

(*Enter R. to him,* EVANGELIST)

EVANGELIST. Wherefore dost thou cry?

CHRISTIAN. Sir, I perceive by the book in my hand that I am condemned to die, and after that to come to judgement, and I find that I am not willing to do the first, nor able to do the second.

EVANGELIST. Why not willing to die, since this life is attended with so many evils?

CHRISTIAN. Because I fear that this burden that is upon my back will sink me lower than the grave.

EVANGELIST. If this be thy condition, why standest thou still?

CHRISTIAN. Because I know not whither to go.

EVANGELIST. Flee from wrath to come.

CHRISTIAN. Whither must I fly?

EVANGELIST. Do you see yonder wicket-gate?

CHRISTIAN. No.

EVANGELIST. Do you see yonder shining light?

CHRISTIAN. I think I do.

EVANGELIST. Keep that light in your eye, and go up directly thereto; so shalt thou see the gate; at which when thou knockest it shall be told thee what thou shalt do.

(*He goes out, L.B.* CHRISTIAN *starts to follow him. His wife and children cry after him, but he heeds them not. She calls up the neighbours*)

CHRISTIAN. Life! Life! Eternal life!

The Pilgrim's Progress

(*Two of the neighbours,* OBSTINATE *and* PLIABLE,
try to hold him back by force)

CHRISTIAN. Neighbours, wherefore are ye come? (*He moves to the front*)

OBSTINATE. To persuade you to go back with us.

CHRISTIAN. That can by no means be; you dwell in the City of Destruction, the place also where I was born; I see it to be so; and dying there sooner or later, you will sink lower than the grave, into a place that burns with fire and brimstone. (*They glance at each other*) Be content, good neighbours, and go along with me.
(*The middle curtains close behind them*)

OBSTINATE. What! And leave our friends and our comforts behind us?

(PLIABLE *laughs*)

CHRISTIAN. Yes, because that *all* which you forsake is not worthy to be compared with a little of that which I am seeking to enjoy; and, if you will go along with me, and hold it, you shall fare as I myself; for there, where I go, is enough and to spare. (*He turns to two loiterers*) Come away, and prove my words.
(*But they run away, crying 'He's mad, he's mad'*)

PLIABLE. What are the things you seek, since you leave all the world to find them?

CHRISTIAN. I seek an inheritance incorruptible, undefiled, and that fadeth not away, and it is laid up in heaven, and safe there to be bestowed on them that diligently seek it. Read it so, if you will, in my book.

OBSTINATE. Tush! Away with your book. Will you go back with us or no?

CHRISTIAN. No, not I, because I have laid my hand to the plough, Obstinate.

OBSTINATE. Come then, neighbour Pliable, let us turn again, and go home without him; there is a company of these

76

crazy-headed coxcombs, that, when they take a fancy by the end, are wiser in their own eyes than seven men that can render a reason.

PLIABLE.   Don't revile; if what the good Christian says is true, the things he looks after are better than ours; my heart inclinest to go with my neighbour.

OBSTINATE.   What! more fools still! Be ruled by me, and go back; who knows whither such a brain-sick fellow will lead you? Go back, go back, and be wise.

CHRISTIAN.   Nay, but do thou come with thy neighbour, Pliable; there are such things to be had which I spoke of, and many more glories besides.

PLIABLE.   Well, neighbour Obstinate, I begin to come to a point; I intend to go along with this good man, and to cast in my lot with him; but, my good companion, do you know the way to this desired place?

CHRISTIAN.   I am directed by a man whose name is Evangelist, to speed me to a little gate that is before me.

PLIABLE.   Come, then, good neighbour, let us be going.

OBSTINATE.   And I will go back to my place. I will be no companion of such misled, fantastical fellows.

(Exit through curtains)

CHRISTIAN.   Come, neighbour Pliable, I am glad you are persuaded to go along with me.

PLIABLE.   Come, neighbour Christian, since there are none but us two here, tell me now further what the things are, and how to be enjoyed, whither we are going.

CHRISTIAN.   I can better conceive of them with my mind, than speak of them with my tongue; but yet, since you are desirous to know, I will read of them in my book.

PLIABLE.   And do you think that the words of your book are certainly true?

77

CHRISTIAN. Yes, verily; for it was made by Him that cannot lie.

PLIABLE. Well said, what things are they?

CHRISTIAN. There is an endless kingdom to be inhabited, and everlasting life to be given us, that we may inhabit that kingdom forever.

PLIABLE. Well said; and what else?

CHRISTIAN. There are crowns of glory to be given us, and garments that will make us shine like the sun in the firmament of heaven!

PLIABLE. This is very pleasant; and what else?

CHRISTIAN. There shall be no more crying, nor sorrow; for He that is owner of the place will wipe all tears from our eyes.

PLIABLE. And what company shall we have there?

CHRISTIAN. There you shall meet with thousands and ten thousands that have gone before us to that place; none of them hurtful, but loving and holy; everyone walking in the sight of God, and standing in His presence with acceptance forever. There we shall see men that by the world were cut in pieces, burnt in flames, eaten of beasts, drowned in the seas, for the love that they bare to the Lord of the place, all well, and clothed with immortality as with a garment.

(*The middle curtains open discovering the Slough of Despond*)

PLIABLE. But are these things to be enjoyed?

CHRISTIAN. The Governor of the country hath recorded that in this book; he will bestow it upon us freely.

PLIABLE. Well, my good companion, glad am I to hear these things; come on, let us mend our pace.

CHRISTIAN. I cannot go so fast as I would, by reason of this burden that is upon my back.

(*They draw near and fall into a 'very miry slough'.*
A low profile of rocks)

PLIABLE. What's this? What's this? We have fallen into a bog, and I am grievously bedaubed with dirt.

CHRISTIAN. Because of my burden I sink deeper into the mire.

PLIABLE. Ah! neighbour Christian, where are you now?

CHRISTIAN. Truly, I do not know.

PLIABLE. And is this the happiness you have told me all this while of? If we have such ill speed at our first setting out, what may we expect betwixt this and our journey's end? May I get out again with my life, you shall possess the brave country alone for me.

(Singing in the distance. PLIABLE *gives a desperate struggle or two, and gets out of the mire on the side next his own house. Exit* PLIABLE)

(Enter HELP, *singing, centre)*

HELP. What do you here?

CHRISTIAN. Sir, I was bid go this way by a man called Evangelist, who directed me also to yonder gate, that I might escape the wrath to come; and as I was going thither I fell in here.

HELP. But why did you not look for the steps?

CHRISTIAN. Fear followed me so hard, that I fled the next way and fell in.

HELP. Then give me thy hand. My name is Help, and I will draw thee out.

*(*CHRISTIAN *does so, and is helped out)*

CHRISTIAN. Sir, wherefore, since over this place is the way from the City of Destruction to yonder gate, is it that this place is not mended, that poor travellers might go thither with more security?

HELP. This miry slough is such a place as cannot be mended, it is the descent whither the scum and filth that attends conviction of sin doth continually run, and therefore it is called the Slough of Despond. It is not the pleasure of the King that this place should remain so bad. His labourers also have, by the direction of His

Majesty's surveyors, been for above these sixteen hundred years employed about this patch of ground, and to my knowledge, here have been swallowed up at least twenty thousand cart-loads, yea, millions of wholesome instructions, that have at all seasons been brought from all places of the King's dominions, and they that can tell, say they are the best materials to make good ground of the place; if so be, it might have been mended, but it is the Slough of Despond still, and so will be when they have done what they can.

(*He sings again. Exit* HELP, *R.F. Curtains close.*

CHRISTIAN *rubs off the mud.* MR WORLDLY WISEMAN *enters in front*)

MR WORLDLY WISEMAN. How, now, good fellow, whither away after this burdened manner?

CHRISTIAN. Sir, I am going to yonder wicket-gate before me; for there, as I am informed, I shall be put into a way to be rid of my heavy burden.

MR WORLDLY WISEMAN. Wilt thou hearken unto me if I Worldly Wiseman of Morality give thee counsel?

CHRISTIAN. If it be good, I will; for I stand in need of good counsel.

MR WORLDLY WISEMAN. I would advise thee, then, that thou with all speed get thyself rid of thy burden; for thou wilt never be settled in thy mind till then.

CHRISTIAN. That is that which I seek, for ever to be rid of this heavy burden; but get it off myself, I cannot; nor is there any man in our country that can take it off my shoulders; therefore am I going this way, as I told you, that I may be rid of my burden.

MR WORLDLY WISEMAN. Who bid thee go this way to be rid of thy burden?

CHRISTIAN. A man that appeared to me to be a very great and honourable person; his name, as I remember, is Evangelist.

MR WORLDLY WISEMAN. I beshrew him for his counsel! There is not a more dangerous and troublesome way in the world than is that unto which he hath directed thee; and that thou shalt find, if thou wilt be ruled by his counsel. (*Laughing maliciously*) Thou hast met with something, as I perceive already; for I see the dirt of the Slough of Despond is upon thee; but that slough is the beginning of the sorrows that do attend those that go on that way. (*Lays a hand on him*) Hear me, I am older than thou; thou art like to meet with, in the way which thou goest, wearisomeness, painfulness, hunger, perils, nakedness, sword, lions, dragons, darkness, and, in a word, death and what not! And why should a man so carelessly cast away himself, by giving heed to a stranger?

CHRISTIAN. Why, sir, this burden upon my back is more terrible to me than are all these things which you have mentioned; I care not what I meet with in the way, if so be I can also meet with deliverance from my burden.

MR WORLDLY WISEMAN. How camest thou by the burden at first?

CHRISTIAN. By reading this book in my hand.

MR WORLDLY WISEMAN. I thought so; and it so happened unto thee as to other weak men, who, meddling with things too high for them, do suddenly fall into thy distractions.

CHRISTIAN. I know what I would obtain; it is ease for my heavy burden.

(*He moves away*)

MR WORLDLY WISEMAN. But why wilt thou seek for ease this way, seeing so many dangers attend it? I could direct thee to the obtaining of what thou desirest, without the dangers; yes, and the remedy is at hand, where instead of those dangers, thou shalt meet with much safety, friendship and content.

CHRISTIAN. Pray, sir, open this secret to me.

MR WORLDLY WISEMAN. Why, in yonder village named Morality —there dwells a gentleman whose name is Legality, a very judicious man, and a man of a very good name, that has skill to help men off with such burdens as thine are from their shoulders; yea, to my knowledge, he hath done a great deal of good this way; ay, and besides, he hath skill to cure those that are somewhat crazed in their wits with their burdens. And if thou art not minded to go back to thy former habitation (as indeed I would not wish thee), thou mayest send for thy wife and children to thee in this village, where there are houses now standing empty, one of which thou mayest have at reasonable rates; provision is there also cheap and good; and that which will make thy life the more happy is, to be sure, there thou shalt live by honest neighbours, in credit and good fashion.

CHRISTIAN. Sir, which is my way to this honest man's house?

MR WORLDLY WISEMAN. Do you see yonder high hill?

CHRISTIAN. Yes, very well.

MR WORLDLY WISEMAN. By that hill you must go, and the first house you come at is his. Give him my name—say, Mr Worldly Wiseman sent you. God speed.

(*Exit* MR WORLDLY WISEMAN, *R.F.*)

CHRISTIAN. I will go to Mr Legality's house for help; but the way seems so high, and my burden seems heavier now than while I was in the way. (*Lightning and thunder*) That was a flash of fire out of the hill. I am afraid I shall be burned.

(EVANGELIST *enters, C., and bars his path*)

EVANGELIST. What dost thou here, Christian?

(CHRISTIAN *does not know what to answer*)

Art thou not the man that I found crying without the walls of the City of Destruction?

CHRISTIAN.   Yes, dear sir, I am the man.

EVANGELIST.   Did I not direct thee the way to the little wicket-gate?

CHRISTIAN.   Yes, dear sir.

EVANGELIST.   How is it, then, thou art so quickly turned aside? For thou art now out of the way.

CHRISTIAN.   I met with a gentleman as soon as I had got over the Slough of Despond, who persuaded me that I might, in the village before me, find a man that could take off my burden.

EVANGELIST.   I will now show thee who it was that deluded thee. The man that met thee is one Worldly Wiseman, and rightly is he so called because he savoureth only the doctrine of this world.

CHRISTIAN.   I am sorry I have hearkened to this man's counsel. But may my sin be forgiven?

EVANGELIST.   Thy sin is very great, for by it thou hast committed two evils; thou hast forsaken the way that is good, to tread in forbidden paths; yet will the man at the gate receive thee, for he hath goodwill for men; only take heed that thou turn not aside again, lest thou perish from the way, when his wrath is kindled but a little. God speed.

(*Exit* EVANGELIST, *C.*)

CHRISTIAN.   On will I go with haste, neither speaking to any man on the way, nor, if any ask me, will I vouchsafe them an answer. I walk like one that is all the while treading on forbidden ground.

(*The middle curtains open, discovering a gate over which is written 'Knock, and it shall be opened unto you'.* CHRISTIAN *looks at the inscription*)

(*He knocks*)

(*A grave person,* GOODWILL, *comes to the gate*)

GOODWILL.   Who is there? Whence come ye? What would ye have?

The Pilgrim's Progress

CHRISTIAN. Here is a poor burdened sinner, journeying towards Mount Zion, that I may be delivered from the wrath to come. And since I am informed that by this gate is the way thither, I would know, sir, if you are willing to let me in?

GOODWILL. I am willing with all my heart. I am Goodwill.

(*He opens the gate. A fleet of arrows is shot. A sound of devilish laughter.* GOODWILL *gives* CHRISTIAN *a pull*)

CHRISTIAN. What means that?

GOODWILL. A little distance from this gate, there is erected a strong castle, of which Beelzebub is the captain; from thence, both he and them that are with him shoot arrows at those that come up to this gate, if haply they may die before they can enter in.

CHRISTIAN. I rejoice and tremble.

GOODWILL. Who directed you hither?

CHRISTIAN. Evangelist bid me come hither, and knock.

(*Close middle curtains*)

GOODWILL. An open door is set before thee, and no man can shut it.

CHRISTIAN. Now I begin to reap the benefits of my hazards. But can you not help me off with my burden that is upon my back? For as yet I have not got rid thereof, nor can I by any means get it off without help.

GOODWILL. As to your burden, be content to bear it, until you come to the place of deliverance; for there it will fall from your back of itself. (*Points*) Up that highway must you go that is fenced on either side by a wall that is called salvation.

(*Exit R.F.*)

(*Open middle curtains, discovering a stone cross above.* CHRISTIAN *moves to it*)

CHRISTIAN. He hath given me rest by his sorrow and life by his death.

(*As he kneels at the foot of the cross his burden rolls off into a hole below. Enter three shining ones*—ANGELS)

CHRISTIAN. Glory to God in the highest.

ANGELS. Peace be to thee.

FIRST ANGEL. Thy sins be forgiven thee.

SECOND ANGEL. Thy rags I take and clothe thee with change of raiment.

THIRD ANGEL. This mark I set upon thy forehead. Take this roll with a seal upon it and look thou givest it in at the Celestial Gate.

ANGELS. Peace be to thee.

(*Tableau*)

CHRISTIAN. Thus far did I come laden with my sin;
Nor could aught ease the grief that I was in
Till I came hither: what a place is this!
Must here be the beginning of my bliss?
Must here the burden fall from off my back?
Must here the strings that bound it to me crack?
Blest cross! blest sepulchre! blest rather be
The man that there was put to shame for me!

(CHRISTIAN *comes forward. After, close middle curtains*)

Now here midway to the top of the hill is a pleasant arbour, made for the refreshing of weary travellers. I will sit down and pull out my roll and read therein to my comfort.

(*He reads, and* SHEPHERD BOY *sings, off*)

SHEPHERD BOY. He that is down needs fear no fall;
He that is low, no pride;
He that is humble, ever shall
Have God to be his guide.

I am content with what I have,
Little be it, or much:
And, Lord, contentment still I crave,
Because thou savest such.

> Fullness to such a burden is,
>   That go on pilgrimage;
> Here little, and hereafter bliss,
>   Is best from age to age.

CHRISTIAN.   I will dare to say that this boy lives a merrier life and wears more of that herb called heartsease in his bosom than he that is clad in silk and velvet.

(He falls asleep. The song continues)
(Enter SHEPHERD BOY, *C.)*

SHEPHERD BOY.   Go to the ant, thou sluggard. Consider her ways, and be wise.

(The SHEPHERD BOY *awakens* CHRISTIAN, *and passes out, R.F.*
CHRISTIAN *starts up, and moves. Enter* TIMOROUS *and* MISTRUST, *L.)*

CHRISTIAN. Sirs, what's the matter? Why run the wrong way?

TIMOROUS.   We were going to the City of Zion, and had got up this difficult place, but the further we go the more danger we meet with. Wherefore we turned and are going back again.

MISTRUST.   Yes, for just before us be a couple of lions in the way, whether sleeping or waking we know not, and we could not think, if we came within reach, but that they would presently pull us to pieces.

CHRISTIAN.   You make me afraid, but whither shall I fly to be safe? If I go back to my own country, that is prepared for fire and brimstone, and I shall certainly perish there. If I can get to the Celestial City I am sure to be in safety there. I must venture to go forward in fear of death and life everlasting beyond it. I will yet go forward.

*(*TIMOROUS *and* MISTRUST *run out.* CHRISTIAN *moves up.*
He feels in his bosom for his scroll)

CHRISTIAN.   My roll! My roll! I have lost my roll!
(He searches around)

Oh, wretched man, that I should sleep in the daytime, and sleep in the midst of difficulty. How many steps have I trod in vain. Yea, now I am like to be benighted, for the day is almost spent...oh, that I had not slept. (*He sees the scroll under the seat he slept upon.*

He hides it in his bosom again)

Thanks be to God for directing my eye to the place where it lay. How nimbly now shall I go up the hill. I must be ware of the lions that affrighted Timorous and Mistrust. These beasts range in the night for their prey, and if they should meet with me in the dark, how should I shift them?

(*The first roar is heard.* CHRISTIAN *stands still.*

Enter WATCHFUL. *Curtains slightly opened*)

WATCHFUL. (*crying to him*) Is thy strength so small? Fear not the lions, for they are chained, and are placed there for trial of faith where it is, and for discovery of those that had none. Keep in the midst of the path, and no hurt shall come to thee.

CHRISTIAN (*entering*) Sir, what house is this? And may I lodge here tonight?

WATCHFUL. This house was built by the Lord of the hill, and he built it for the relief and security of pilgrims. Whence come you, and whither are you going?

CHRISTIAN. I am going to Mount Zion; but because the sun is now set, I desire, if I may, to lodge here tonight.

WATCHFUL. What is your name?

CHRISTIAN. My name is now Christian, but my name at the first was Graceless.

WATCHFUL. But how doth it happen that you come so late? The sun is set.

CHRISTIAN. I had been here sooner, but that I slept in the arbour that stands on the hillside.

WATCHFUL. I will call out the virgins of this place, who will, if they like your talk, bring you in, according to the rules of the house. Come, Prudence, Piety and Charity!

(*He calls for* PRUDENCE, PIETY *and* CHARITY, *who come*)

CHARITY. Come in, thou blessed of the Lord. This house was built by the Lord of the hill, on purpose to entertain such pilgrims in.

(*He bows his head, and follows them through the curtains, which open discovering a terrace*)

PIETY. Come, good Christian, since we have been so loving to you, to receive you in our house this night, let us, if perhaps we may better ourselves thereby, talk with you of all things that have happened to you on your pilgrimage.

CHRISTIAN. With a very good will, and I am glad that you are so well disposed.

PRUDENCE. And what is it that makes you so desirous to go to Mount Zion?

CHRISTIAN. Why, there I hope to see him alive that did hang on the cross; and there I hope to be rid of all those things that to this day are in me an annoyance to me; there, they say, there is no death; and there I shall dwell with such company as I like best. For to tell you the truth I love him, because I was by him eased of my burden; and I am weary of my inward sickness. I would fain be where I shall die no more, and with the company that shall continually cry, 'Holy! Holy! Holy!'.

CHARITY. Have you a family? Are you a married man?

CHRISTIAN. I have a wife and four small children.

CHARITY. And why did you not bring them along with you?

CHRISTIAN. Why, my wife was afraid of losing this world, and my children were given to the foolish delights of

The Pilgrim's Progress

youth; so, what by one thing and what by another, they left me to wander in this manner alone.

CHARITY. But did you not, with your vain life, damp all that you used by way of persuasion to bring them away with you?

CHRISTIAN. Indeed I cannot commend my life. Yet this I can say, I was very wary of giving them occasion, by an unseemly action, to make them averse to going on pilgrimage.

(Meanwhile WATCHFUL *and* DISCRETION *have brought in a table furnished with wine and food)*

PRUDENCE. Supper is ready. Come, Piety, Charity, and Discretion. Let us sit down to meat.

(They move. PIETY *murmurs a grace)*

PIETY. The Lord of this hill, who built this house for the relief and security of pilgrims, was a great warrior, and hath fought with and slain him that hath power over death.

CHARITY. But not without great danger to himself, Piety, which made me love him the more. Look towards the South. At a great distance are the Delectable Mountains.

PIETY. It is a most pleasant country, beautified with woods, vineyards, fruits of all sorts, flowers also, with springs and fountains, very delectable to behold.

CHRISTIAN. What is it called?

PIETY. It is Immanuel's land, and it is as common as this Hill for all pilgrims.

CHARITY. And when thou comest there, thou mayest see to the Gate of the Celestial City, as the shepherds that live there will make appear.

PRUDENCE. Let us commit ourselves to the Lord for protection and betake ourselves to rest.

(They move)

PIETY. And on the morrow, before thou settest forward, we will go to the armoury for such furniture as the Lord hath provided for Pilgrims.

PRUDENCE. There shalt thou be harnessed with what is proof lest thou meet with assaults on the way...and now, O Pilgrim, thou shalt be laid in a large upper chamber whose window opens towards the sun rising.

PIETY. The name of the chamber is Peace.

(They go out. Close curtains)

INTERVAL

ACT II

(Front scene (middle curtains closed).
Enter CHRISTIAN, *fully armed, and porter,* WATCHFUL.)

CHRISTIAN. Saw you any pilgrims pass by?

WATCHFUL. Yes.

CHRISTIAN. Pray, did you know him?

WATCHFUL. I asked him his name, and he told me it was Faithful.

CHRISTIAN. Oh, I know him; he is my townsman, my near neighbour; he comes from the place where I was born. How far do you think he may be before?

WATCHFUL. He is got by this time below the hill.

CHRISTIAN. Well, good Watchful, the Lord be with thee, and add to all thy blessings much increase, for the kindness that thou hast showed to me.

(Enter DISCRETION, PIETY, CHARITY *and* PRUDENCE, *R.F.)*

CHRISTIAN. As it was difficult coming up, so far as I can see, it is dangerous going down.

PRUDENCE. Yes, so it is, for it is a hard matter for a man to go down into the valley of Humiliation, as thou art now, and to catch no slip by the way.

CHARITY. Take for your refreshment this loaf of bread, this bottle of wine, and a cluster of raisins.

PIETY. The Lord be with thee, and guide thy feet.

(*They wave adieu, and exeunt R.F.*
Middle curtains open revealing APOLLYON)

APOLLYON. (*to* CHRISTIAN, *re-entering, L. upper*) Whence come you? And whither are you bound?

CHRISTIAN. I am come from the City of Destruction, which is the place of all evil, and am going to the City of Zion.

APOLLYON. By this I perceive thou art one of my subjects, for all that country is mine, and I am the prince and god of it. How is it, then, that thou hast run away from thy king? Were it not that I hope thou mayest do me more service, I would strike thee now at one blow to the ground.

CHRISTIAN. I was born indeed in your dominions, but your service was hard, and your wages such as a man could not live on, for the wages of sin is death.

APOLLYON. There is no prince that will thus lightly lose his subjects, neither will I as yet lose thee.

CHRISTIAN. But I have let myself to another, even to the King of Princes.

APOLLYON. Thou hast done in this, according to the proverb, 'Changed a bad for a worse'.

CHRISTIAN. I have given him my faith, and sworn my allegiance to him; how, then, can I go back from this, and not be hanged as a traitor?

APOLLYON. Thou didst the same to me, and yet I am willing to pass by all, if now thou wilt yet turn again and go back.

CHRISTIAN. O thou destroying Apollyon! To speak truth, I like his service, his wages, his servants, his government, his company and his country better than thine; and,

therefore, leave off to persuade me further; I am his
servant, and I will follow him.　　(*Moves*)

APOLLYON.　Thou hast already been unfaithful in thy service to
him; and how dost thou think to receive wages of him?
Thou art inwardly desirous of vain-glory in all thou
sayest or doest.

CHRISTIAN.　All this is true, and much more which thou hast left
out; but the Prince whom I serve and honour is
merciful, and ready to forgive.

APOLLYON.　I am an enemy to this Prince; I hate his person, his
laws, and people; I am come out on purpose to withstand thee.

CHRISTIAN.　Apollyon, beware what you do; for I am in the
King's highway, the way of holiness; therefore, take
heed to yourself.

APOLLYON.　I am void of fear in this matter; prepare thyself to
die; for I swear by my infernal den, that thou shalt go
no further; here will I spill thy soul.

　　(*They fight. Music.* CHRISTIAN *is almost spent*)

APOLLYON.　I am sure of thee now.

CHRISTIAN.　Rejoice not against me, O mine enemy; when I fall
I shall arise.

(*He gives* APOLLYON *a deadly thrust which makes him fall backward*)
Nay, in all these things we are more than conquerors
through him that loved us.

　　(*Exit* APOLLYON, *U.R.* CHRISTIAN *faints.*
　　Enter, centre, an ANGEL. *Music*)

ANGEL.　These leaves from the tree of life shall heal thy wounds
immediately.

(*They are applied to* CHRISTIAN'S *wounds, and he is healed. He eats
the bread and drinks the wine that he has been given. The* ANGEL *goes.
Close middle curtains on* ANGEL. CHRISTIAN *still lies faint. Enter two
men*)

CHRISTIAN. Whither are you going?

FIRST MAN. Back! Back! And we would have you so do too, if either life or peace is prized by you.

CHRISTIAN. Why, what's the matter?

SECOND MAN. Matter! We were going that way as you are going, and went as far as we durst; and indeed we were almost past coming back; for had we gone a little further, we had not been here to bring the news to thee.

CHRISTIAN. But what have you met with?

FIRST MAN. Why, we were almost in the Valley of the Shadow of Death; but that, by good hap, we looked before us, and saw the danger before we came to it.

CHRISTIAN. But what have you seen?

SECOND MAN. Seen! Why, the valley itself, which is as dark as pitch; we also saw there the hobgoblins, satyrs and dragons of the pit; we heard also in that valley a continual howling and yelling, as of a people under unutterable misery, who sat there bound in affliction and irons; and over that valley hang the discouraging clouds of confusion.

FIRST MAN. Death also doth always spread his wings over it. In a word, it is every whit dreadful, being utterly without order.

CHRISTIAN. Yet I perceive notwithstanding that this is my way to the desired haven.

MEN. Be it thy way; we will not choose it for ours.

(*Exeunt* MEN. *Darkness comes. Middle curtains open slowly.*
CHRISTIAN *struggles forward with the sword drawn, then falls again*)

CHRISTIAN. Oh, Lord, I beseech thee, deliver my soul!

(*Smoke and flames across his path; thunder and lightning.
Fiendish laughter*)

CHRISTIAN. I will walk in the strength of the Lord God! Though I walk through the Valley of the Shadow of Death,

I will fear no evil, for thou art with me; and glad will I be because there be some who fear God in this valley as well as myself. For I perceive God is with them, though in that dark and dismal state; and why not with me, though by reason of the impediment that attends this place, I cannot perceive it. Could I but overtake them, I should have company by and by. He hath turned the shadow of death into the morning.

(*Dawn. Music and light. He kneels*)

He discovereth deep things out of darkness, and bringeth out to light the shadow of death. His candle shineth upon my head, and by his light I walk through darkness.

(CHRISTIAN *rises*)

(*Enter* FAITHFUL, *above L.*)

CHRISTIAN.   Ho! Ho! Soho! Stay, and I will be your companion!

(FAITHFUL *looks around*)

Stay, stay, till I come up with you.

FAITHFUL.   No, I am set out upon my life, and the avenger of blood is behind me.

CHRISTIAN.   My honoured and well-beloved brother, Faithful, I am glad that I have overtaken thee; and that God has so tempered our spirits, that we can walk as companions in this so pleasant a path.

FAITHFUL.   I had thought, dear friend, to have had your company quite from our town; but you did get the start of me, wherefore I was forced to come thus much of the way alone.

CHRISTIAN.   How long did you stay in the City of Destruction, before you set out after me on your pilgrimage?

FAITHFUL.   Till I could stay no longer; for there was a great talk presently after you were gone out, that our city would be burned to the ground.

CHRISTIAN.   What! Did our neighbours talk so?

FAITHFUL. Yes, it was for a while in everybody's mouth.

CHRISTIAN. Did you hear no talk of neighbour Pliable?

FAITHFUL. Yes, Christian, I heard that he followed you till he came at the Slough of Despond, where, as some said, he fell in; but he would not be known to have so done; but I am sure he was soundly bedabbled with that kind of dirt.

CHRISTIAN. And what said the neighbours to him?

FAITHFUL. He hath, since his going back, been had greatly in derision, and that among all sorts of people; some do mock and despise him; and scarce will any set him on work. He is now seven times worse than if he had never gone out of the city.

CHRISTIAN. But why should they be so set against him, since they also despise the way that he forsook?

FAITHFUL. Oh, they say, hang him, he is a turncoat! He was not true to his profession.

CHRISTIAN. Had you no talk with him before you came out?

FAITHFUL. I met him once in the streets, but he leered away on the other side, as one ashamed of what he had done; so I spake not to him.

CHRISTIAN. Well, neighbour Faithful, let us leave him, and talk of things that more immediately concern ourselves.

(*Enter* TALKATIVE, *R.F.*)

FAITHFUL. Friend, whither away? Are you going to the heavenly country?

TALKATIVE. I am going to the same place.

FAITHFUL. That is well; then I may hope we have your good company.

TALKATIVE. With a very good will will I be your companion.

FAITHFUL. Come on, then, and let us go together, and let us spend our time in discoursing of things that are profitable.

TALKATIVE.   What thing is so pleasant, and what so profitable, as to *talk* of the things of God?

FAITHFUL.   But, by your leave, heavenly knowledge is the gift of God; no man attaineth to it by human industry, or only by the talking of it.

TALKATIVE.   All this I know very well; for a man can receive nothing, except it be given him from heaven; all is of grace, not of works. I would give you a hundred scriptures for the confirmation of this.

FAITHFUL.   Well, then, what is that one thing that we shall at this time found our discourse upon?

TALKATIVE.   What you will. I will talk of things heavenly, or things earthly; things moral or things evangelical; things sacred or things profane; things past or things to come; things foreign or things at home; things more essential or things circumstantial; provided that all be done to our profit.

(*He pulls out a little book and studies it*)

FAITHFUL.   (*Stepping towards* CHRISTIAN *who is walking apart, softly*) What a brave companion have we got! Surely this man will make a very excellent pilgrim.

CHRISTIAN.   This man with whom you are so taken, will beguile with that tongue of his twenty of them that know him not.

FAITHFUL.   Do you know him, then?

CHRISTIAN.   Know him! Yes, better than he knows himself.

FAITHFUL.   Pray, what is he?

CHRISTIAN.   His name is Talkative: (*They laugh*) he dwelleth in our town. I wonder that you should be a stranger to him.

FAITHFUL.   Whose son is he? And whereabouts does he dwell?

CHRISTIAN.   He is the son of one Say-Well; he dwelt in Prating Row and is known of all that are acquainted with him by the name of Talkative in Prating Row; and, notwithstanding his fine tongue, he is but a sorry fellow.

FAITHFUL. Say you so? Then am I in this man greatly deceived?

CHRISTIAN. (*with rising indignation*) Deceived! You may be sure of it. He talketh of prayer, of repentance, of faith and of new birth, but he knows but only to talk of them. I have been in his family and have observed him both at home and abroad. His house is as empty of religion as the white of an egg is of savour. He thinks that hearing and saying will make a good Christian, and thus he deceiveth his own soul.

FAITHFUL. Well, I was not so fond of his company at first, but I am sick of it now. What shall we do to be rid of him?

CHRISTIAN. Take my advice, and do as I bid you, and you shall find that he will soon be sick of your company too, except God shall touch his heart and turn it.

FAITHFUL. What would you have me to do?

CHRISTIAN. Why, go to him, and enter into some serious discourse about the power of religion; and ask him plainly, whether this thing be set up in his heart, house or conversation.

FAITHFUL. (*stepping towards* TALKATIVE) Come, what cheer? How is it now?

TALKATIVE. Thank you, well. I thought we should have had a great deal of talk by this time.

FAITHFUL. Well, if you will, we will fall to it now; and since you left it with me to state the question, let it be this—How doth the saving grace of God discover itself when it is in the heart of man?

TALKATIVE. I perceive, then, that our talk must be about the power of things. Well, it is a very good question, and

I shall be willing to answer you. And take my answer
in brief, thus: First, where the Grace of God is in the
heart it causeth there a great outcry against sin.
Secondly....

FAITHFUL. Nay, hold, let us consider of one at once. I think you
should rather say, It shows itself by inclining the soul
to abhor sin.

TALKATIVE. Why, what difference is there between crying out
against and abhorring of sin?

FAITHFUL. Oh, a great deal. But what is the second thing
whereby you would prove a discovery of a work of
grace in the heart?

TALKATIVE. Great knowledge of gospel mysteries.

FAITHFUL. Knowledge, great knowledge, may be obtained in the
mysteries of the gospel, and yet no work of grace in
the soul. To know is a thing that pleaseth talkers and
boasters; but to *do* is that which pleaseth God.

TALKATIVE. You lie at the catch. This is not for edification.

FAITHFUL. Well, if you please, propound another sign how this
work of grace discovereth itself where it is.

TALKATIVE. Not I, for I see we shall not agree. This kind of
discourse I did not expect, nor am I disposed to give
answer to such questions, because I count not myself
bound thereto. I refuse to make you my judge. Since
you are ready to judge so rashly as you do, I cannot but
conclude you are some peevish or melancholy man,
not fit to be discoursed with, and so adieu.

(*Exit* TALKATIVE, *L.F.*)

CHRISTIAN. (*coming up to* FAITHFUL) I told you how it would
happen: he had rather leave your company than reform
his life.

(*Middle curtains open, discovering four pedlars and eight country folk
who dance, while some players set up their stage, and two booths*

are set up with wares. A crowd collects, men crying wares, numerous devils crying:)

DEVILS. Merchant Devils! We buy souls for gold—souls for gold. We buy souls for gold (etc.).

MOUNTEBANK *(upon the little stage at back, shouts)* Beelzebub, Apollyon and Legion Limited, vanity good! Houses, lands, trades, places, honours, preferments, titles, lusts, pleasures, bawds, wives, lives, blood, bodies, gold, silver, pearls, precious stones and what not.

(He repeats it, beating a drum. The Nobility arrive, among whom LORD LUXURY, LADY LECHERY, SIR HAVING GREEDY, MISTRESS HEADY, LORD CARNAL DELIGHT, LADY VAIN GLORY, MASTER LOVE LUST, MISTRESS MALICE. *They dance a pavane. The medley of sounds increases)*

LADY FEIGNING. *(to* MRS NO-GOOD) Who be these outlandish men, clothed in raiment so diverse from our own, Mrs No-Good?

MRS NO-GOOD. They be fools or bedlams, lady, so may come to our Fair with nothing to sell! *(Laughter)*

LADY FEIGNING. And with no great desire to buy, for, look you, they turn their eyes upwards as if their trade and traffic were in Heaven.

MRS NO-GOOD. They are barbarians, My Lady Feigning, and speak the language of Canaan. Mr Pickthank, invite them to Vanity Fair.

MR PICKTHANK. I will speak to them, you shall hear their answer. *(To* CHRISTIAN) What will you buy?

CHRISTIAN. We buy the truth. *(Murmurs)*

LADY FEIGNING. It is dangerous to let such fools have their liberty, Superstition.

SEVERAL. It is dangerous to the Commonweal.

MR SUPERSTITION. Whence come ye? Whither could ye go?

CHRISTIAN. We are pilgrims and strangers in the world, going to our own country, which is the Heavenly Jerusalem.

MR ENVY. (*indignantly*) These be the vilest of men, Mr Highmind, who bring all good custom to nought.

MR HIGHMIND. For ought I can see, Envy, the men are quiet and sober, intending nobody any harm, and there be many that trade in this fair more worthy to be abused than these.

(*General hubbub*)

MR ENVY. Have them in chains as an example or tenor to others, lest any should follow them and trade against our city.

HOPEFUL. Nay, save these men, for they seem good and true, and of fair speech, seeing that they buy the truth.

(*More noise and fighting. Enter the* WATCH, *U.L.*)

MR PICKTHANK. Hale these men before the judges as disturbers of the peace.

CROWD. To the Prison—the Stocks—the Cage—bring them before the Judge—let them be bound—call the jury together.

MR SUPERSTITION. To whom will you now commit yourselves?

FAITHFUL. To the all-wise disposal of him that ruleth all things, and with good content.

(*Uproar. Enter the* JUDGE, LORD HATEGOOD, *from the back*)

CLERK. Silence, for his Honour, Lord Hategood of Vanity Fair!

(*The crowd hurriedly arrange a court, the following, on the right being the Jury. Their names are called.* 1, MR BLINDMAN, *foreman;* 2, MR NO-GOOD; 3, MR MALICE; 4, MASTER LOVE LUST; 5, MR LIVELOOSE; 6, MR HEADY; 7, MR HIGHMIND; 8, MR ENMITY; 9, MR LIAR; 10, MR CRUELTY; 11, MR HATE-LIGHT; 12, MR IMPLACABLE)

CLERK. Silence in Court.

JUDGE. Read the Indictment.

CLERK. These men, Christian and Faithful, of the Heavenly Jerusalem, are enemies to and disturbers of our trade. They have made commotions and divisions in the town and have won a party to their own most dangerous opinions in contempt of the law of our prince.

JUDGE. What defence hast thou to make?

FAITHFUL. I have only set myself against that which hath set itself against Him that is higher than the highest. And as for disturbance, I make none, being myself a man of peace.

MR ENVY. My Lord, I have known this man a long time and will attest upon my oath, before this honourable bench that he is—

JUDGE. Hold! Give him his oath. (*They swear him*)

MR ENVY. My Lord, this man, notwithstanding his plausible name, is one of the vilest men in our country. He neither regardeth prince nor people nor law nor custom, but doth all that he can to possess all men with certain of his disloyal notions, which he in the general calls principles of faith and holiness. And, in particular, I heard him once myself affirm that Christianity and the customs of our town of Vanity were diametrically opposite and could not be reconciled. By which saying, my Lord, doth he at once not only condemn all our laudable doings, but ourselves in the doing of them.

JUDGE. Hast thou any more to say?

MR ENVY. My Lord, I could say much more, only I would not be tedious to the court.

(SUPERSTITION *is called, and bade to look upon the prisoner. Then he is sworn*)

MR SUPERSTITION. My Lord, I have no great acquaintance with this man, nor do I desire to have further knowledge of him; however, this I do know, that he is a very pestilent fellow, from some discourse that I had with him in this town; for then, talking with him, I heard him say that our religion was nought, and one by which a man could by no means please God. Which sayings of his, my

Lord, your Lordship very well knows, what necessarily hence will follow, to wit, that we still do worship in vain, are yet in our sins, and finally shall be damned, and this is that which I have to say.

MR PICKTHANK. My Lord, and you gentlemen all, this fellow I have known of a long time, and have heard him speak things that ought not to be, for he hath railed on our noble prince, Beelzebub, and hath spoken contemptibly of his honourable friends, whose names are the Lord Old Man, the Lord Carnal Delight, the Lord Luxurious, the Lord Desire of Vain Glory, my old Lord Lechery, Sir Having Greedy, with all the rest of our nobility; and he hath said, moreover, that if all men were of his mind, if possible, there is not one of these noblemen should have any longer a being in this town. Besides, he hath not been afraid to rail on you, my Lord, who art now appointed to be his Judge, calling you an ungodly villain, with many other such like vilifying terms with which he hath bespattered most of the gentry of our town.

JUDGE. Thou runagate, heretic and traitor, hast thou heard what these honourable gentlemen have witnessed against thee?

FAITHFUL. May I speak a few words in my own defence?

JUDGE. Sirrah, sirrah, thou deservest to live no longer, but to be slain immediately upon the place; yet that all men may see our gentleness towards thee, let us hear, what thou, vile runagate, hast to say.

FAITHFUL. I say, then, in answer to what Mr Envy hath spoken, I never said ought but this, that what rule or laws or customs or people were flat against the word of God, were diametrically opposite to Christianity. If I have said amiss in this, correct me of my error, and I am ready here before you to make my recantation. As to the second, to wit Mr Superstition and his charge against me, I said this only, that in the worship of God there is required a divine faith, but there can be no divine faith without a divine revelation of the will of God. As to what Mr Pickthank hath said, I say that the prince of this town, with all the rabblement, his attendants, by this gentleman named, are more fit for a being in Hell than in this town and country, and so the Lord have mercy upon me.

JUDGE. Gentlemen of the jury, you see this man about whom so great an uproar hath been made in this town. You have also heard what these worthy gentlemen have witnessed against him. Also you have heard his reply and confession. It lieth now in your breasts to hang him or to save his life; but yet I think meet to instruct you into our law. (*Clears his throat*) There was an act made in the days of Pharaoh the Great, servant to our Prince, that lest those of a contrary religion should multiply and grow too strong for him, their males should be thrown into the river. There was also an act made in the days of Nebuchadnezzar the Great, another of his servants, that whosoever should not fall down and worship his golden image should be thrown into a fiery furnace. There was also an act made in the days of Darius that whoso for some time called upon any god but him, should be cast into the Lions' den. Now the substance of these laws this rebel has broken, not only in thought, but also in word and deed, which must therefore needs be intolerable. For that of Pharaoh, his law was made upon a supposition to prevent mischief, no crime being yet apparent, but here is a crime apparent. For the second and third you see he disputeth against our religion, and for the treason he hath confessed, he deserveth to die the death.

The Pilgrim's Progress

MR BLINDMAN (*foreman of the jury*).   I see clearly that this man is a heretic.

MR NO-GOOD.   Away with such a fellow from the earth.

MR MALICE.   Aye, for I hate the very looks of him.

MASTER LOVE LUST.   I could never endure him.

MR LIVELOOSE.   For he would always be condemning my way.

MR HEAVY.   Hang him.

MR HIGHMIND.   A sorry scrub.

MR ENMITY.   My heart riseth against him.

MR LIAR.   He is a rogue.

MR CRUELTY.   Hanging is too good for him.

MR HATELIGHT.   Let us despatch him out of the way.

MR IMPLACABLE.   Might I have all the world given to me, I could not be reconciled to him. Let us forthwith bring him in guilty of death.

ALL.   Guilty of Death.

JUDGE.   (*pronouncing sentence*) With all due expediency let the man Faithful be bound to the stake and burnt to ashes.

   (*Hubbub*)

HOPEFUL.   (*to* CHRISTIAN *aside*) Take my cloak and hide thee. Hereafter upon the way I will join thee, for they have forgotten thee in their anxiety to do ill to thy brother. Nay, speak not, but wrap my cloak about thee.

(FAITHFUL *is bound to the stake. The executioner seems to set fire to the faggots. The shining ones appear behind* FAITHFUL *ready to take his soul. Tableau*)

(*Middle curtains close*)

INTERVAL
(*if necessary; otherwise straight on*)

ACT III

(Front Scene, middle curtains closed. Enter CHRISTIAN *and*
HOPEFUL, *L.F.)*

HOPEFUL. I will join myself unto you, and entering into a
brotherly covenant will be your companion.

CHRISTIAN. Thus, one died to bear testimony to the truth, and
another rises out of his ashes, to be a companion with
Christian in his pilgrimage.

HOPEFUL. There were many more of the men in the Fair, that
would take their time and follow after.

(Enter BYENDS, *R.F.)*

CHRISTIAN. What countryman, sir? And how far go you this way?

BYENDS. From the town of Fair-speech, and I am going to the
Celestial City.

CHRISTIAN. Fair-speech! Is there any good that lives there?

BYENDS. Yes, I hope.

CHRISTIAN. Pray, sir, what may I call you?

BYENDS. I am a stranger to you, and you to me; if you be going
this way, I shall be glad of your company; if not, I
must be content.

CHRISTIAN. This town of Fair-speech I have heard of, and, as I
remember, they say it is a wealthy place.

BYENDS. Yes, I will assure you that it is, and I have very many
rich kindred there.

CHRISTIAN. Pray, who are your kindred there, if a man may be
so bold?

BYENDS. Almost the whole town; and in particular, my Lord
Turn-about, my Lord Time-server, my Lord Fair-
speech (from whose ancestors, that town first took its
name); also Mr Smooth-man, Mr Facing-both-ways,
Mr Anything, and the parson of our parish, Mr Two-
tongues, was my mother's own brother by father's

side; and to tell you the truth, I am become a gentleman of good quality; yet my grandfather was but a waterman, looking one way and rowing another, and I got most of my estate by the same occupation.

CHRISTIAN. Are you a married man?

BYENDS. Yes, and my wife is a very virtuous woman, the daughter of a virtuous woman; she was my Lady Feigning's daughter, therefore she came of a very honourable family, and is arrived to such a pitch of breeding that she knows how to carry it all, even to prince and peasant. It is true we somewhat differ in religion from those of the stricter sort, yet but in two points; first, we never strive against wind and tide; secondly, we are always most zealous when religion goes in his silver slippers; we love much to walk with him in the streets, if the sun shines, and the people applaud him.

CHRISTIAN. It runs in my mind that this is one Byends of Fairspeech; and if it be he, we have as very a knave in our company as dwelleth in all these parts.

HOPEFUL. Ask him; methinks, he should not be ashamed of his name.

CHRISTIAN. Sir, you talk as if you knew something more than all the world doth, and if I take not my mark amiss, I deem I have half a guess of you. Is not your name Mr Byends of Fair-speech?

BYENDS. This is not my name, but indeed it is a nickname that is given me by some that cannot abide me; and I must be content to hear it as a reproach, as other good men have borne theirs before me.

CHRISTIAN. But did you never give an occasion to men to call you by this name?

BYENDS. Never, never! The worst that ever I did to give them an occasion to give me this name was, that I had always the luck to jump in my judgement with the present way of times, whatever it was, and my chance was to get thereby.

CHRISTIAN. I thought indeed that you were the man that I heard of; and to tell you what I think, I fear this name belongs to you more properly than you are willing we should think it doth.

BYENDS. You shall find me a fair company-keeper, if you will still admit me your associate.

CHRISTIAN. If you will go with us, you must go against the wind and tide; the which, I perceive, is against your opinion; you must also own religion in his rags, as well as when in his silver slippers; and stand by him, too, when bound in irons, as well as when he walketh the streets with applause.

BYENDS. You must not impose, nor lord it over my faith; leave me to my liberty, and let me go with you.

CHRISTIAN. Not a step further, unless you will do in what I propound.

(*Middle curtains open, discovering* MR HOLD-THE-WORLD, MR MONEY-LOVE *and* MR SAVE-ALL)

BYENDS. I shall never desert my old principles, since they are harmless and profitable. If I may not go with you, I must do as I did before you overtook me, even go by myself, until some overtake me that will be glad of my company.

(*Exit* CHRISTIAN *and* HOPEFUL, *R.F.*)

MR MONEY-LOVE. (*coming forward and bowing low*) Who are they upon the road before us?

BYENDS. (*bowing*) They are a couple of far countrymen, Money-Love, that, after their mode, are going on pilgrimage.

MR MONEY-LOVE. Alas! Why did they not stay, that we might have their good company? For they, and we, and you, sir, I hope, are all going on a pilgrimage.

BYENDS. We are so indeed; but the men before us are so rigid, that let a man be never so godly, yet if he jumps not with them in all things, they thrust him quite out of their company.

MR MONEY-LOVE. That is bad, but we read of some that are righteous overmuch, and such men's rigidness prevails with them to judge and condemn all but themselves. But I pray, what and how many were the things wherein you differed?

BYENDS. Why, they, after their head-strong manner, conclude that it is a duty to rush on their journey all weathers; and I am waiting for wind and tide. They are for religion when in rags and contempt; but I am for him when he walks in his golden slippers, in the sunshine and with applause. What do you say, Hold-the-World?

MR HOLD-THE-WORLD. Aye, and hold you here still (*bowing*), good Mr Byends, for, for my part, I can count him but a fool, that, having the liberty to keep what he has, shall be so unwise as to lose it. Abraham and Solomon grew rich in religion, and Job says that a good man shall lay up gold as dust. Eh, Save-all?

MR SAVE-ALL. (*bowing*) I think that we are all agreed in this matter, indeed, for he that believes neither scripture nor reason, neither knows his own liberty, nor seeks his own safety.

BYENDS. (*calling to* CHRISTIAN) Hold there, my friends; Mr Hold-the-World, Mr Save-all and Mr Money-Love are all going on pilgrimage (*Re-enter* CHRISTIAN *and* HOPEFUL) and would propose this question.

MR HOLD-THE-WORLD. (*to* CHRISTIAN *and* HOPEFUL) Suppose a man, a minister, or a tradesman, etc., should have an advantage lie before him, to get the good blessings of this life, yet so as that he can by no means come by

The Pilgrim's Progress

them except, in appearance at least, he becomes extraordinarily zealous in some points of religion that he meddled not with before; may he not use these means to attain his end, and yet be a right honest man?

CHRISTIAN. Even a babe in religion may answer ten thousand such questions. For if it be unlawful to follow Christ for loaves (as it is in the sixth of John), how much more abominable is it to make of him and religion a stalking horse, to get and enjoy the world! Nor do we find any other than heathens, hypocrites, devils and witches, that are of this opinion. Neither will it hold out of my mind, but that that man that takes up religion for the world will throw away religion for the world; for so surely as Judas resigned the world in becoming religious, so surely did he also sell religion and his Master for the same thing. To answer the question, therefore, affirmatively, as I perceive you have done, and to accept as authentic such answer, is both heathenish, hypocritical and devilish; and your reward will be according to your works.

(They stand staring, one upon another. CHRISTIAN comes forward)
(Close curtains)

CHRISTIAN. *(to HOPEFUL)* If these men stand before the sentence of men, what will they do with the sentence of God? And if they are mute when dealt with by vessels of clay, what will they do when they shall be rebuked by the flames of a devouring fire?

(Enter DEMAS, L.F.)

DEMAS. Ho! Turn aside hither, and I will show you a thing.

CHRISTIAN. Who are you?

DEMAS. Demas of Hill Lucre.

CHRISTIAN. What thing so deserving as to turn out of the way to see it?

DEMAS. Here is a silver mine, and some digging in it for treasure.

109

If you will come, with a little pains you may fairly provide for yourselves.

HOPEFUL.   Let us go see.

CHRISTIAN.   Not I. I have heard of this place before now, and how many have there been slain; and besides that, treasure is a snare to those that seek it; for it hindereth them in their pilgrimage. Is not the place dangerous? Hath it not hindered many in their pilgrimage?

DEMAS.   Not very dangerous except to those that are careless.

CHRISTIAN.   (*to* HOPEFUL) Let us not stir a step, but still keep on our way.

HOPEFUL.   I will warrant you, when Byends comes up, if he hath the same invitation as we, he will turn in thither to see.

CHRISTIAN.   No doubt thereof, for his principles lead him that way, and a hundred to one but he dies there. The ground is deceitful and many have been maimed there, and could not, to their dying day, be their own men again.

(*Enter R.F.* BYENDS *and his companions*)

DEMAS.   Ho, turn aside hither and I will show you a thing. A delicate path called Ease to the Hill called Lucre!

BYENDS.   At the first beck!

(*All go out L., except* CHRISTIAN *and* HOPEFUL)

CHRISTIAN.   Oh, Hopeful, whether they fall over the brink or whether they are smothered by the damps in the mire, I fear we shall never see Byends and his companions again in the way.

HOPEFUL.   So these do take up in this world, and no further go.

CHRISTIAN.   If this meadow lieth along by our wayside, let us go over into it. Here is the easiest going; come, good Hopeful, and let us go over.

HOPEFUL.   But how if this path should lead us out of the way?

CHRISTIAN. That is not like. Look, doth it not go along by the
    wayside?

EVANGELIST'S *voice* (*off*). Set thine heart toward the highway,
    even the way which thou wantest; turn again.

    (*Enter C.* GIANT DESPAIR *and his wife* DIFFIDENCE)

GIANT DESPAIR. Diffidence! Who be these sturdy rogues that
    have trespassed on me by trampling in my grounds?

DIFFIDENCE. They be pilgrims or such like. Now, Despair, I
    counsel you to beat them without mercy with your
    grievous crabtree cudgel—and then bid them to make
    away with themselves as thou hast done many a time
    and oft.

GIANT DESPAIR. Why standest thou, fool-hardy, by Doubting
    Castle, yea, even against the gate of Giant Despair.
    (*He seizes them*) Shall I not drive you to a very dark
    dungeon, nasty and stinking to the spirit? There shall
    ye forthwith make an end of yourselves seeing that
    life is attended with so much bitterness, and if thou
    dost not it shall be the worse for thee.

    (*Exeunt L.* GIANT DESPAIR, CHRISTIAN *and* HOPEFUL.
        DIFFIDENCE *remains*)

DIFFIDENCE. (*to herself*) Now doth my husband beat them and
    bind them? I would to God that he hath not one of his
    fits that restraineth his hand, but that he may despatch
    them immediately. Meanwhile I will think upon what
    counsel I can give him. (*Re-enter L.* GIANT DESPAIR)
    Are they despatched?

GIANT DESPAIR. They are sturdy rogues; they choose rather to
    bear all hardship than to make away with themselves.

DIFFIDENCE. Take them into the castle yard tomorrow, and show
    them the bones and skulls of those thou hast already
    despatched, and make them believe that ere a week
    comes to an end thou also wilt tear them in pieces as
    thou hast done their fellows before. I fear they live in

hope that someone will relieve them, or that they have some picklocks about them by the means of which they hope to escape.

GIANT DESPAIR. Sayest thou so, my dear? I will, therefore, search them in the morning.

(*Exeunt* GIANT DESPAIR *and* DIFFIDENCE)

(*Open middle curtains, discovering a dungeon, and* CHRISTIAN *and* HOPEFUL *in chains*)

CHRISTIAN. Brother, what shall we do? The life that we now lead is miserable. For my part I know not whether it is best to live thus, or to die out of hand. 'My soul chooseth strangling rather than life', and the grave is more easy for me than this dungeon. Shall we be ruled by the giant?

HOPEFUL. The lord of the country to which we are going hath said Thou shalt do no murder, no, not to another man's person. Much more, then, are we forbidden to take his counsel and kill ourselves. Besides, he that kills another can but commit murder upon his body, but for one to kill himself is to kill body and soul at once. My brother, let us be patient, and endure a while. The time may come that may give us a happy release; but let us not be our own murderers.

(*Enter* GIANT DESPAIR, *a little on L.*)

GIANT DESPAIR. Look through the lattice yonder. There are the bones of those that once were pilgrims, but they trespassed on my ground, as you have done; so when I thought fit I tore them in pieces, and so, within ten days, I will do you, if you have not already killed yourselves.

(GIANT DESPAIR *withdraws again*)

HOPEFUL. My brother, rememberest thou not how valiant thou hast been heretofore? Apollyon could not crush thee, nor could all that thou didst hear or see in the Valley of the Shadow of Death. What hardship, terror,

amazement hast thou already gone through! And art
thou now nothing but fear! Wherefore let us bear up
with patience as well as we can.

> Who would true valour see,
>   Let him come hither;
> One here will constant be,
>   Come wind, come weather.
> There's no discouragement
> Shall make him once relent
> His first avowed intent
>   To be a pilgrim.
>
> Who so beset him round
>   With dismal stories
> Do but themselves confound—
>   His strength the more is;
> No lion can him fright,
> He'll with a giant fight;
> But he will have a right
>   To be a pilgrim.
>
> Hobgoblin nor foul fiend
>   Can daunt his spirit;
> He knows he at the end
>   Shall life inherit.
> Then fancies flee away,
> He'll fear not what men say;
> He'll labour night and day
>   To be a pilgrim.

CHRISTIAN. What a fool I am thus to lie in a stinking dungeon,
when I may as well walk at liberty! I have a key in my
bosom called Promise, that will, I am persuaded, open
any lock in Doubting Castle.

HOPEFUL. That is good news, brother, pluck it out of thy
bosom and try.

The Pilgrim's Progress

(CHRISTIAN *pulls it out of his bosom, and begins to try the dungeon door, the bolts of which, as he turns the key, give way, and the door flies open with ease.* CHRISTIAN *and* HOPEFUL *both come out as the middle curtains close. Enter L.F.* 3 SHEPHERDS *and* A BOY)

CHRISTIAN.   That lock went damnable hard!

(CHRISTIAN *and* HOPEFUL *come forward with* SHEPHERDS)

CHRISTIAN.   Whose delectable mountains are these, shepherds?

HOPEFUL.   And whose be the sheep that feed upon them?

FIRST SHEPHERD.   These mountains are Immanuel's land, and they are within sight of his City, and the sheep also are his, and he laid down his life for them.

CHRISTIAN.   Is this the way to the Celestial City?

FIRST SHEPHERD.   You are just in your way.

CHRISTIAN.   How far is it thither?

FIRST SHEPHERD.   Too far for any but those that shall get thither indeed.

CHRISTIAN.   Is the way safe or dangerous?

SECOND SHEPHERD.   Safe for those for whom it is to be safe, but the transgressors shall fall therein.

CHRISTIAN.   Is there in this place any relief for pilgrims that are weary and faint in the way?

FIRST SHEPHERD.   The Lord of these mountains hath given us a charge not to be forgetful to entertain strangers. Therefore the good of the place is before you.

THIRD SHEPHERD.   Welcome to Delectable Mountains.

SECOND SHEPHERD.   Our names are Knowledge, Experience, Watchful and Sincere. Come to our tents, and partake of that which is ready at present.

FIRST SHEPHERD.   We would that ye should stay here awhile, to be acquainted with us, and yet more to solace yourselves with the good of these Delectable Mountains.

CHRISTIAN.   We are content to stay because it is very late.

HOPEFUL.   We have need to cry to the Strong for strength.

114

THIRD-SHEPHERD.   Ay, and you will have need to use it, when you have it, too.

SECOND SHEPHERD.   Let us here show to the pilgrims the gates of the Celestial City if they have skill to look through our perspective glass.   (*Hands* CHRISTIAN *glass*)

CHRISTIAN.   (*giving it to* HOPEFUL)   My hand is unsteady; I can see nothing.

HOPEFUL.   I think I see a gate.

(CHRISTIAN *and* HOPEFUL *lie by the Shepherds' fire. The dawn begins*)

HOPEFUL.   (*musing*) The air here is very sweet and pleasant; yea, we hear continually the singing of birds; the flower appears in the earth, and the voice of the turtle is heard in the land.

(*The middle curtains open, discovering the three* ANGELS)

CHRISTIAN.   (*dreaming*) Behold the shining ones as the bride-groom rejoiceth over the bride, so did their God rejoice over them.

SHINING ONES.   Say ye to the daughter of Zion,
              Behold thy salvation cometh,
              Behold his reward is with him.

CHRISTIAN   (*shrinking*) I am weak with longing.

(*The* GARDENER *approaches them from L.F.*)

HOPEFUL.   (*to* GARDENER) Whose goodly vineyards and gardens are these, O Gardener?

THE GARDENER.   They are the King's and are planted here for his own delight, and also for the solace of the pilgrims; therefore, refresh yourselves with these dainties.

(*They take grapes and wine cup. Exit* GARDENER, *L.F.*)

SHINING ONES.   Whence come ye?

HOPEFUL.   From the City of Destruction by the way of Vanity Fair.

SHINING ONES.   All temptations and dangers have ye passed, but ye have two difficulties more to meet with and then you are within the City.

CHRISTIAN.   Wilt thou go along with us?

SHINING ONES.   You must obtain it by your own faith. Behold, betwixt us is the river of death, there is no bridge to go over and the river is very deep.

HOPEFUL.   What shall we do?

SHINING ONES.   You must go through or you cannot come to the gate.

CHRISTIAN.   Is there no other way to the gate?

SHINING ONES.   Yes, but there hath not any save two, to wit, Enoch and Elijah, been permitted to tread that patch, since the foundations of the world, nor shall until the last trumpet shall sound.

CHRISTIAN.   Are the waters all of a depth?

SHINING ONES.   No; yet we cannot help you in that case, for you shall find it deeper or shallower as you believe in the King of the place.

(Close middle curtains. Darkness)

CHRISTIAN.   *(dying)* I sink in deep waters; the billows are over my head. All his waves go over me.

HOPEFUL.   *(moves to him)* Be of good cheer, my brother. I feel the bottom, and it is good.

CHRISTIAN.   Ah, my friend, the sorrows of death have encompassed me about; I shall not see the land that flows with milk and honey.   *(He sinks)*

HOPEFUL.   Brother, I see the gate, and the men standing by to receive us.

CHRISTIAN.   It is for you. It is for you they wait. You have been hopeful ever since I knew you.

HOPEFUL.   And so have you.

CHRISTIAN.   Ah, brother, surely if I was right, he would now arise and help me, but for my sins he hath brought me into this snare and left me.

HOPEFUL.   Be of good cheer; Jesus Christ maketh thee whole.

CHRISTIAN.   Saying 'when thou passest through the waters I will be with thee, and through the rivers, they shall not overflow thee'.

(*He dies. Music. The middle curtains open*)

(*The* SHINING ONES *appear above. They help them up*)

SHINING ONES.   We are the ministering spirits sent forth for those that shall be heirs of Salvation. You are going now to the Paradise of God, wherein you shall see the tree of Life, and eat of the never-fading fruits thereof; you shall have white robes given you, and your walk and talk shall be every day with the King, even all the days of eternity.

(*Two* TRUMPETERS *come R. and L.* CHORUS OF ANGELS *follows*)

TRUMPETERS.   Blessed are they that are called to the Marriage Supper of the Lamb.

CHORUS OF ANGELS.   Blessed are they that do his commandments, that they may have right to the tree of Life and may enter in through the gates into the City.   (*The bells ring*)

SHINING ONES.   Enter ye into the joy of your Lord.

CHORUS OF ANGELS.   Blessings and honour and glory and power, be unto him that sitteth upon the throne, and unto the Lamb for ever and ever.

OMNES.   (*as outer curtain closes*)
> Holy, Holy, Holy is the Lord.
> Holy, Holy, Holy is the Lord.

(BUNYAN *comes forward*)

BUNYAN.   What of my dross thou findest here, be bold
To throw away, but yet preserve the gold;
What if my gold be wrapped up in ore?
None throws away the apple for the core.
But if thou shalt cast all away as vain,
I know not but 'twill make me dream again.
(BUNYAN *goes out*)

THE END

Death in the Tree

('Der Dot im Stock')

This play tells the
same tale as Chaucer's
Pardoner's Tale

THE CHARACTERS

| THE HERMIT | THE BULLY |
| THE CYNIC | THE COWARD |

Ther cam a privee theef, men clepeth Deeth,
That in this contree al the people sleeth....

The Pardoner's Tale, CHAUCER

(Scene: A wood. Entrances right and left. There is a stump of a dead tree left, and from one bough hangs a dead crow. A broken branch, large enough to serve as a seat, lies on the right. The ground is covered with dead leaves. It is an evening in late autumn. Enter the HERMIT, right. He is a very old and grey-bearded man in a grey robe. He leans on his staff and mutters as he walks.)

HERMIT. *Quis est fides?*
 Quod non vides.
 Quis est spes?
 Quod non habes.
 Quis est caritas?
  Rara raritas.[1]
How many more tired hungry days
Must I, poor hermit, go from place to place?
I have renounced the world, journeying
On foot from day to day, to serve my heavenly King.
Caring nothing for the world's joys, I
Seek neither possessions, nor felicity,
But with deliberate hunger, observances, and tears
I seek to pass away my earthly years.

---

[1] What is faith? What we do not see. What is hope? What we do not have. What is charity? The rarest rarity.

Death in the Tree

Ah, but today I'm weary; and maybe
I could take time to rest under this tree—
To listen to the songs of praise
Raised by the wood birds here, gaze
On all these works of God, who, wise and good,
Made them, gave them their song, gives them their food.

(He strikes the tree with his staff)

Ha? That's strange: this trunk sounds hollow:
From this drum-beat what episode will follow?

(He looks in the tree)

A sack of money? Ah, God help me then:
Let gold not lead me into temptation.
Solomon, the wisest man, so scripture says,
Wrote of how all our earthly vanities—
Hoards, robes, coins, property—all bring disaster.
Yet all men strive to gather vaster
Estates, by honest or dishonest means,
Grasping their way through lies, greedy for gains
So I shall now do well to give you up, discovery—
Lie there in hiding.   *(Drops the bag into the tree)*
              Not a penny!
But as if Death were in this tree,
I'll put what miles I can between that sack and me.

(He goes off, but hesitates and returns)

And yet, perhaps, supposing I meet a poor
Beggarman? With a handful I could restore
His health, allay his hunger: then maybe
Gold is no root of evil, but a means to mercy?

(He takes the sack out again and opens it)

God help me! What strange old coins are these
Engraved with savage symbols, effigies
Of forgotten emperors? Now conscience strive,
Guide me! What use is treasure trove
As alms? No beggar could spend it. What use is it to me?
Back in the bag. Evil, hide in your rotten tree!

Death in the Tree

(He flings the money back into the tree stump and crosses himself. Exit quickly left. Enter THE BULLY, THE CYNIC, *and* THE COWARD, *right.* THE BULLY *and* THE CYNIC *sit down and begin to clean their weapons.* THE COWARD, *ill at ease, remains standing.)*

BULLY. Well it's long enough I reckon since
　　　Anything good came our way. Chance
　　　Gave us some good game last week, sport,
　　　Remember we near got nabbed, what,
　　　The day we met Sir Sheriff and his party?
　　　Might have been under lock and key
　　　Now, if it weren't for this overgrown old wood.

COWARD. Might have been worse than that: cold,
　　　At a rope's end. Stiff. Blood for blood,
　　　That's what they'll take from us. Dead,
　　　That's what we'd be by now. It don't bear thinking of.

CYNIC. Well, we know that, don't we, well enough?
　　　Know what kind of 'reform' they'd offer us?
　　　So, we don't choose to take their reward:
　　　Besides they outnumber us, Coward!
　　　And here we are, alive—while they're still chasing shadows.

COWARD. They've got us marked. Likenesses put around:
　　　Sometime they'll get us, Cynic, run us to ground.
　　　We're hunted killers: men without friends,
　　　All three of us. Men with identical ends:
　　　One verdict'll do for us: one sentence: then it's U.P., up.
　　　　　　(He makes as if being hanged)
　　　They'll have no mercy on us. Let's get away.
　　　　　　(He makes as if to run. BULLY *holds him back)*

BULLY.　　　　　　　　　　　　　　　No! Stop!
　　　You cowardly six feet of gallows-meat:
　　　An outlaw's life is short, but ain't it sweet?
　　　Havn't you heard the tale they put around
　　　How my old dame—she's an old witch, bound

By the Devil himself to work black spells (why,
She can turn churches upside down by the wink of an eye!)
She's put a spell on us: so we can't be took
While she still goes through the proper rigmarole in the
book.

So you'll stay out of jail while there's influence
There to protect you. Besides, when the last bad chance
Trips you up: when you're an old lag, strength's
Gone—done every crime in the book—your length's
Stretched out, in bed, or on the gallows: tell me, mate,
Does it matter which? You can say to Fate,
'*Exeo*': here I go. Can't you stand a few seconds' pain?
You only die once, man: they can't hang you over again.

COWARD.　　Ah, but does it all end there? All very well
If misery came to an end with the funeral bell:
But every soul that's sinned must go to the fire,
And burn everlastingly.

CYNIC.　　　　　　　　The Bible's a liar:
No soul's lost: if they won't have it up in Heaven
Old Hornie catches it then. And, why, he even
Goes out of his way to find cool spots,
Ways to soften the burns, or loosen the bolts,
So's everyone can bear it. What would be the use
Of millions suffering down in Hell, endless
Pain for ever, eh? No point at all. You mark me well—
The parsons all exaggerate the pains of Hell.
For why? Because they know they'd be out of work
If we all thought Hell could be easy. We'd shirk,
Have a good time, break all the rules:
So they frighten the poor with sulphur, silly fools.

COWARD.　　Is this the time to blaspheme? Deny
What everybody believes? Just when our destiny
Catches up with us? Our arrest,
That's certain soon. Then we will be confessed:

An hour or two later, there we'll stand
Right before God, in front of his face. And
Then what'll you say? Ask for grace?
Mercy from our last Judge? Say the place
Doesn't exist? Your disbelief won't do you much good
There, will it?

BULLY.   Listen, Coward—
Let me tell you there's no heaven, nor Hell,
No Devil, no God neither. Good and Evil
Are all the same. And when you're dead
You're done for: dead as mutton, dead
As any animal is, after you've cut its throat.

COWARD.   I'll tell you what you are: you're a man of sin,
Worse than savages, for ever they maintain
Some sort of belief in a kind of after-state.

CYNIC.   Kid's talk, all this about life and death.
When hunger drives like it drives us, don't waste your
                                          breath:
It's bread you want, here, now, not after-lives:
That stuff's all empty wind, all about heaven and Hell,
Souls flying up in the sky when they toll the bell!
                    (*He laughs derisively*)
I tell you what, though: if a merchant
Came by with a wallet, he'd want
To know about it all sooner than we do: 'cause I'd drill
A bullet through him, string him up, and he'll
Soon see if there's an after-life.

BULLY.   Sure!
Now you're talking plain.

CYNIC.   We'd take his cares
From him for ever: send him up those heavenly stairs!

COWARD. (*Looking off*)   There! Look through the trees: there's
                                          someone.
An old man. Looks scared. Keeps breaking into a run.

125

CYNIC.　Yes! And keeps looking back as if he had
　　　Sovereigns sewn in his seams.

BULLY.　　　　　　　　　　　　　This old running Dad
　　　We'll chase, and cut his throat, and hide
　　　His carcass in the bushes. Then one of us can ride
　　　To town with the loot—bring back
　　　Drink, grub. My belly has gone all slack:
　　　I could do with a meal. Coming this way. Hide, quick,
　　　Here by this tree. Now, you club, and I'll stick.

COWARD.　Why! It's the Hermit of the Woods! The holy
　　　　　　　　　　　　　　　　　　　　　　man!

　　　Carries no money, gold, nothing anyone
　　　Can make use of. Lives in a poor way,
　　　Gives up all worldly goods. A bad day
　　　When we murder them that's poorer than we are.
　　　Put your sword up. He's old and weak, half-bare...

CYNIC.　What's all this pious talk? What's got you?
　　　Seen the light suddenly? What are you going to do?
　　　Change your trade to be a parson—Mr Goody-goody?
　　　Are you going to faint, eh, when you see him lie all
　　　　　　　　　　　　　　　　　　　　　　bloody?

　　　Better get in a monastery, man, and hide,
　　　And leave the woods where only brave men ride.

BULLY.　In the Devil's name keep quiet: He's in the glade.

COWARD.　This'll be the end of us. An omen.

CYNIC.　　　　　　　　　　　　　　　Why, you're afraid.

　　　　　(*They hide.* HERMIT *enters, looking furtively over
　　　　　his shoulder at the oak.* BULLY *stands in his way*)

BULLY.　Where are you making for, old man? Ha? Hey?
　　　Why have you left your cell? Gone a bit astray?

CYNIC.　What d'you keep cringing for? And looking back?
　　　Did you see Old Nick behind you on the track?

126

Death in the Tree

HERMIT. I have seen Death here in the Hollow Tree:
I have heard Death among these paths, pursuing me.
So let me go in safety on my way:
And, strangers, leave that tree to the wind's sigh.
(*The wind moans, the dead crow swings. A pause, and then
the* CYNIC *draws his sword*)

CYNIC. Why, in the twinkling of an eye
We can show you that much, easily.
Before you leave this solitary place
We'll treat you to another look at old Death's face.

HERMIT. (*Falling on his knees*) If you are Christian men, please
let me go
My way in peace. Surely you must know
That if you take a life your crime
Shall bring you retribution, at some time,
Man's or God's. (BULLY *and* CYNIC *seize him*) God sees what
you do now
To one who has no guilt. May Christ forgive you.
(*They drag him out right. A heavy blow is heard followed by a cry.*
COWARD *sinks down on the log in terror*)

COWARD. He said 'May Christ forgive you'. But we forfeit
grace:
This murder means the end of us, now, in this place.
(*Re-enter* BULLY *and* CYNIC *wiping their swords*)

BULLY. Number eighteen! Notch it up, mates: this wood
Contains them all, left to rot in the bracken here, for
good.

CYNIC. Peace be to the mortal bones of our old friend
Whose dreadful warnings brought him a good end.
Here, where's the way in? Let's see if old John Death
Lurks in the tree, having just took the breath
Out of old Grandad. Or Death's relations, eh?
Skulking down in the dark of this hollow old oak tree.
(*He looks into the stump and pulls out the bag*)

Lord in Heaven. Strike me down! Here's fruit:
Not Death, mate, neither. Gold. Lovely sparkling loot!
A thousand florins, I should say. Well, strike me dead!

BULLY. Old Grandad put it there and all, I bet: got rid
Of worldly care, eh? Thought he'd leave it just
In case he got fed up with hermitting. Christ
Save the poor old looney fellow's soul. Chink!
Do you hear it chink?

CYNIC. Like music, man. And food and drink
That's what this means for us. Why, this bag of death
Will buy us back our nourishment and health.
Good old Dad, to think of his poor starving brothers.

BULLY. One of us now must take some of this to town, while
the others
Stay here and guard the rest, and buy some grub and
booze.
And when we've filled our empty stomachs, we'll work
out who's
To have what shares in what. Right? Here!
Salvation, freedom, security, wealth—what more
Could we want now? (*Runs his hands through the coins*)

COWARD. But which of us shall go?
They're looking for us in the town.

BULLY. Throw lots.

CYNIC. Go on then, throw.

BULLY. Three!

CYNIC. Five!

COWARD. Two! (*The others laugh*)
Why did it fall like that?
I was the one who wanted least to go.

BULLY. (*Giving him coins*) Take these, and get
Meat, bread, and wine without delay. Stay outside the wall
Where the stalls are: don't go in town at all,

For if you're caught, well, we shall have no dinner
And next time we see you, you'll be hanging up, you
                                                sinner,
On the town gate. Then this tree's inside
Will have been full of Death, like Grandad said.

COWARD.   All right. I'll go. Though I can't keep a muscle
Still, with fear. I shake as each bough creaks, at every
                                                rustle
Death's after me. Evil's somewhere, in the dark
Shadows of this wood. Or in the town. Or on the track
After us, like he said, the poor old hermit.   (*Exit*)

BULLY. (*Shouting after him*)   Quick!

CYNIC.   He slinks off like a dog with its tail down:
Afraid of a bit of stabbing or knocking down:
One day he'll split on us, mate, he'll betray
His fond companions. Then we'll have had our day.

BULLY.   That's what I've been thinking, friend, and I
Suggest we hide when he comes back, and then....
          (*Makes a gesture of violence*)
CYNIC.   Why?
What? Do him in?

BULLY.   Liquidate him. Traitors, man,
Must take their medicine. He might blurt out a plan,
Squeak to the beadles. Once and for all we'd free
Ourselves of that. One stab. And you and me
Needn't have no more fear. How's that? What say?

CYNIC.   That's a good scheme. And furthermore—thereby
We'd have a little bigger bit out of the pie.

BULLY.   How's that?

CYNIC.   Obvious. A thousand florins into two
Goes half a thousand each.

BULLY.   That's right, mate. Yes, that's
                                                true.

CYNIC.   If he was here we'd only get three ton and thirty three.

BULLY.   You're a genius you are, man. Here, swear, with me,
You'll help do him in.

CYNIC. (*Shakes* BULLY's *hand*)   Soon as we've got the grub off
him, of course.
Don't want to dirty that!

BULLY. (*They laugh*)   No mess!

BOTH. (*Raising their right hands*)   And here we swear to strike
together.

CYNIC.   When murdering's in hand you find out whether
A man's a true man. And now I've another scheme....

BULLY.   Another? Come on now, let's have it. In this game
We share everything, don't we? Come on, friend.
(BULLY *is uneasy, thinking that perhaps he is being out-witted*)

CYNIC.   Old greybeard, old Grandad, who just had his end,
And stands up there now before God. I wonder
If the cunning old fellow kept some more, under
His shirt, eh? Some sovereigns sewn in his seams?
Didn't think of that, did you? Come on.

BULLY. (*Superstitiously*)   Sometimes....
Suppose we went down there now and found...nothing....

CYNIC.   Well, what have you got to lose? Everything
We find is clear profit for us, isn't it, anyway?

BULLY.   There won't be nothing on him, that's what I say.

CYNIC.   Then let him keep his nothing: it's all one.

BULLY.   Supposing we find some writing or charms what he'd
done,
Poison, or mumbo-jumbo, in a black book, spells—
I've heard these old hermits work up some of hell's
Tricks for their enemies. If it's true, isn't it best
To let a wizard's corpse alone, not start feeling under his
vest?

CYNIC.   Spells! Wizards! You've 'heard'! Why, you're scared:
    Well, you stay here on your knees, while I pull up his beard.
        (*Exit*)

BULLY.   No, I can't stay here alone. I'll come and share
    Anything you find: loot, plans—and the fear....
        (*Exit*)

        (COWARD *enters left*)

COWARD.   What? Gone? Where've they gone? Fled?
    Couldn't stand hanging about so near the dead.
    (*He sees the bag*)
    Bad Conscience got them, did it, at last?
    And here's the bag, untouched, still tied up fast
    As when I left. Yes, but they're hungry too, they'll be
    Back soon, thinking of meat and the wine I've brought
                            with me.
    They'll join me. Ha! Ha! But I'll not join them. No! Oh, no!
    I'll eat. But I'll not be putting my lips to you,
    Mistress bottle. A hasty meal for them, a drink, and then—
    Curtains. I've put a poison in the wine:
    A good wine. Bottle of Death that is. And so
    The world shall benefit from this, in a single blow—
    Two murderers removed, so that any man
    Walking in these dark woods, or a wandering woman
    Can walk safe paths. All of a sudden: gulp!
    Then they'll be gone. And then I'll mend my ways, help
    Hermits, poor people, with this gold, and go
    To foreign places, living as best I know
    How to save my poor soul, penance and prayers
    To win grace and pardon, in my later years.
    And then perhaps I can come to escape God's wrath,
    Hellfire, and the eternal aftermath.
    (*He crouches by the tree.* CYNIC *and* BULLY *return*)

CYNIC.   There's our man, colleague. Come now, quick,
    Cut his throat like a pig's while I do the stabbing trick.

Death in the Tree

BULLY. (*Seizes* COWARD)   Rogue! Why have you been away
so long?

Got lost, eh? Got into town? Took the wrong
Turning—to the narks perhaps? Unloaded
Information, did you? And was you rewarded
For talking, eh? Hope to keep your own white skin
All in one piece, eh?

CYNIC. (*Puts his knife to* COWARD'S *throat*)   Let all the beadles in
the world come in—

And they'll not save you now: nor take us.
Before we smell far off a man in uniform, no fuss,
But sneaks and double-crossers disappears.

COWARD.   I never spoke to no one, never saw no one at all.
Wait. Don't. There's something that I ought to tell...
   (*They kill him and throw him into the bushes*)

CYNIC.   There! That's what we do to sneaks and traitors,
chum,

Send them off smart and quick to Kingdom Come.
Give us a hand—he can lie with Grandad there,
All sleeping healthy in the open air.
   (*They drag* COWARD'S *body out, and return*)
Now we can sit down here in peace, and eat
And wash away all care with a glass of sweet
Red wine. Cheers, brother! Here's good health.
Here's the first instalment of our new-found wealth.

BULLY.   Whether this here is Death or not, it tastes all right,
And we can live like this for years to come! Good appetite!
   (*They sing*)
I went to church and the priest said Evil
Always comes with money. Well, what the devil—
As I went out he was standing at the door
Holding out his box: what the devil did he want it for?

CYNIC.   I met a wise man, and he said Death
Always came with money. Well, he got short of breath.

But when we searched his baggage he had a thousand or
more—
Couldn't take it with him. What did he want it for?

BULLY. (*lugubriously*) We got gold. Found it in a tree.
Know what to do with ours. Slack, eat, be free.
But what about our chum? The one who ain't no more?
What good did it do him? What did he want it for?

Now we'll divide the loot, shall we?
Come on.

CYNIC.　　　　Then we'll get away from this tree
And out of this wood, to town, and live it up, friend, hot,
To the smart hotels, the girls, the Good Time, all the lot!

BULLY. (*suddenly struck by the poison*) Ah! Here, help me,
friend. I'm caught
By a paralysing agony all along my gut.

CYNIC. It's got me too. My heart is bursting, tight
Bands across my chest. I'm shaking like in fright,
And it's all going dark, man, dark, and dying out....

BULLY. He did it. Ratted. Poisoned us—too late,
We're goners, certain. He hoped to get the loot.
I'm done for. Legs have gone now. Death in the tree.
This is our fate.

(Dies. The gold scatters)

CYNIC. All right. So the old man in his grey hood spoke the
truth
When he said this bag of coins was full of Death.
We're only mortal. And I've spent my life
Making out there's no conscience, moral strife,
Nor sin. Well, now Time with the glass and scythe
Takes hold of me, I can hear my long-forgotten faith
Despairing. Clears the mind. And what it says is this:
I go
To face my Maker. And what else I know

133

Death in the Tree

Is that I haven't a hope in Hell of grace.
And here comes a great hot flame from the very place,
And burns me through, like sheets of molten gold
Charring me black, in Hell's dense smoke-clouds rolled...
(*Dies. The wind moans. The dead crow swings.
A few dead leaves fall from the tree*)

EPILOGUE

(*Spoken by* HANS SACHS)

The Christian moral is that worldly gain
May occupy your soul, and fatally. 'Twas plain
The message of St Paul: the root
Of human evil is the love of loot.
We live a little here, then leave all at the gate:
Better live well than die striving to accumulate.
'What's "yours" you can't take with you.' This simple
                                                paradox
We've taken freely from our ancient author, old Hans Sachs.

THE END

Falstaff at Gadshill

Falstaff at Gadshill

THE CHARACTERS

HENRY, PRINCE OF WALES

SIR JOHN FALSTAFF   POINTZ   GADSHILL

PETO   BARDOLPH   MISTRESS QUICKLY

SHERIFF   DRAWERS   CARRIERS

TRAVELLERS, ETC.

SCENE I

London. A tavern

FALSTAFF.   How, Hal, what time of day is it lad?

HENRY.   Thou art so fat-witted, with drinking of old sack and unbuttoning thee after dinner, and sleeping upon benches after noon, that thou hast forgotten to demand that truly which thou wouldst truly know. What the devil hast thou to do with the time of day? Unless hours were cups of sack, and the blessed sun himself a fair hot wench in flame-coloured taffeta. I see no reason why thou shouldst be so superfluous to demand the time of day.

FALSTAFF.   Indeed, you come near me now, Hal; for we that take purses go by the moon and the seven stars, and not by Phoebus. And I prithee, sweet wag, when thou art king—as God save thy Grace—majesty I should say, for grace thou wilt have none,—

HENRY.   What, none?

FALSTAFF.   No, by my troth, not so much as will serve to be prologue to an egg and butter.

HENRY.   Well, how then? Come, roundly, roundly.

FALSTAFF.   Marry, then, sweet wag, when thou art king, let not us that are squires of the night's body be called thieves of the day's beauty; and let men say we be men of good government, being governed, as the sea is, by our noble and chaste mistress the moon, under whose countenance we steal. And is not my hostess of the tavern a most sweet wench?

HENRY.   Why, what have I to do with my hostess of the tavern?

FALSTAFF.   Well, thou hast called her to a reckoning many a time and oft.

HENRY.   Did I ever call for thee to pay thy part?

FALSTAFF.   No, I'll give thee thy due, thou hast paid all there.

HENRY.   Yea, and elsewhere, as far as my coin would stretch; and where it would not I have used my credit.

FALSTAFF.   Yea and used it, that, were it not here apparent that thou art Heir apparent...but, I prithee, sweet wag, shall there be gallows standing in England when thou art King? and resolution thus fobb'd as it is with the rusty curb of old father antick the law? Do not thou, when thou art king, hang a thief.

HENRY.   No; thou shalt.

FALSTAFF.   Shall I? O rare! By the Lord, I'll be a brave judge. But Hal, I prithee, trouble me no more with vanity, I would to God thou and I knew where a commodity of good names were to be bought. An old lord of the Council rated me the other day in the street about you, sir,—but I mark'd him not; and yet he talks very wisely, —but I regarded him not; and yet he talked very wisely, and in the street too.

HENRY.   Thou didst well; for wisdom cries out in the street and no man regards it.

FALSTAFF.   O, thou hast damnable iteration, and art, indeed, able to corrupt a saint. Thou hast done much harm upon me, Hal, God forgive thee for it! Before I knew thee, Hal, I knew nothing; and now am I, if a man should speak truly, little better than one of the wicked. I must give over this life, and I will give it over; By the Lord, an I do not, I am a villain; I'll be damned for never a King's son in Christendom.

HENRY.   Where shall we take a purse tomorrow, Jack?

FALSTAFF.   Zounds, where thou wilt, lad: I'll make one: an I do not, call me villain, and baffle me.

HENRY.   I see a good amendment of life in thee, from praying to pursetaking.

(Enter POINTZ)

FALSTAFF.   Well, Hal, 'tis my vocation, Hal; 'tis no sin for a man to labour in his vocation.—Pointz!—Now we shall know if Gadshill have set a match. O, if man were to be saved by merit what hope in hell were hot enough for him? This is the most omnipotent villain that ever cried 'stand' to a true man.

HENRY.   Good morrow, Ned.

POINTZ.   Good morrow, sweet Hal. What says Sir John Sack and Sugar?

HENRY.   He will give the devil his due.

POINTZ.   But my lads, my lads, tomorrow morning, by four o'clock, early at Gadshill! There are pilgrims going to Canterbury with rich offerings, and traders riding to London with fat purses: I have vizards for you all; you have horses for yourselves. Gadshill lies tonight in Rochester: I have bespoke supper tomorrow night in Eastcheap: we may do it as secure as sleep. If you will go I will stuff your purses full of crowns; if you will not, tarry at home and be hang'd.

FALSTAFF.   Hear ye, Yedward: if I tarry at home and go not, I'll hang you for going.

POINTZ. Will you, chops?

FALSTAFF.   Hal, wilt thou make one?

HENRY.   Who, I rob? I a thief? Not I by my faith.

FALSTAFF.   There's neither honesty, manhood nor good fellow-
ship in thee, nor thou camest not of the blood royal, if
thou darest not stand for ten shillings.

HENRY.   Well, then, once in my days I'll be a madcap.

FALSTAFF.   Why, that's well said.

HENRY.   Well, come what will, I'll tarry at home.

FALSTAFF.   By the Lord, I'll be a traitor, then, when thou art
king.

HENRY.   I care not.

POINTZ.   Sir John, I prithee, leave the prince and me alone; I will
lay him down such reasons for this adventure, that he
shall go.

FALSTAFF.   Well, God give thee the spirit of persuasion, that the
true prince may, for recreation sake, prove a false thief.
Farewell: you shall find me in Eastcheap.

HENRY.   Farewell, thou latter spring! Farewell, All-hallown
summer!

(*Exit* FALSTAFF)

POINTZ.   Nay, my good sweet honey lord, ride with us tomorrow,
I have a jest to execute that I cannot manage alone.
Falstaff, Bardolph, Peto and Gadshill shall rob those
men that we have already waylaid; yourself and I
will not be there; and when they have the booty, if
you and I do not rob them cut this head from my
shoulders.

HENRY.   But how shall we part with them in setting forth?

POINTZ.   Why, we will set forth before or after them, and appoint
them a place of meeting, wherein it is our pleasure to

fall; and then they will adventure upon the exploit themselves; which they shall have no sooner achieved but we'll set upon 'em.

HENRY. Yea, but 'tis like that they will know us by our horses, by our habits, and by every other appointment, to be ourselves.

POINTZ. Tut! Our horses they shall not see—I'll tie them in the wood; our vizards we will change after we leave them, and, sirrah, I have cases of buckram for the nonce, to inmask our noted outward garments.

HENRY. Yea, but I doubt they will be too hard for us.

POINTZ. Well, for two of them, I know them to be as true-bred cowards as ever turn'd back; and for the third, if he fight longer than he sees reason, I'll forswear arms. The virtue of this jest will be, the incomprehensible lies that this same fat rogue will tell us when we meet at supper: how thirty, at least, he fought with; what wards, what blows, what extremities he endured....

HENRY. Well, I will go with thee: provide us all things necessary, and meet me tonight in Eastcheap; there I'll sup. Farewell.

POINTZ. Farewell, my lord.   (*Exit*)

HENRY. I know you all, and will awhile uphold
The unyok'd humour of your idleness:
Yet herein will I imitate the sun,
Who doth permit the base contagious clouds
To smother up his beauty from the world,
That when he please again to be himself,
Being wanted, he may be more wonder'd at.

SCENE II

The road by Gadshill

(*Enter* PRINCE HENRY *and* POINTZ:
BARDOLPH *and* PETO *at some distance*)

POINTZ.   Come, shelter, shelter, I have removed Falstaff's horse,
and he frets like a gummed velvet.

HENRY.   Stand close.   (*They hide*)

(*Enter* FALSTAFF)

FALSTAFF.   Pointz! Pointz and be hang'd! Pointz!

HENRY. (*coming forward*)   Peace, ye fat-kidney'd rascal! What a
brawling dost thou keep!

FALSTAFF.   Where's Pointz, Hal?

HENRY.   He is walk'd up to the top of the hill: I'll go seek him.
(*Retires*)

FALSTAFF.   I am accurst to rob in that thief's company: the rascal
hath removed my horse, and tied him I know not where.
If I travel but four foot by the squier further afoot I
shall break my wind.... Eight yards of uneven ground
is three score and ten miles afoot with me; and the
stony-hearted villains know it well enough: a plague
upon it, when thieves cannot be true to one another!
(*They whistle*) Whew!—A plague upon you all! Give
me my horse, you rogues; give me my horse, and
be hang'd!

HENRY.   Peace, ye fat-guts! lie down; lay thine ear close to
the ground, and list if thou canst hear the tread of
travellers.

FALSTAFF.   Have you any levers to lift me up again, being down?
'Sblood, I'll not bear mine own flesh so far afoot again
for all coin in thy father's exchequer. What a plague
mean ye to colt me thus?

HENRY.   Thou liest: thou art not colt, thou art uncolted.

FALSTAFF. I prithee, good Prince Hal, help me to my horse, good king's son.

HENRY. Out, ye rogue, shall I be your ostler?

FALSTAFF. Go, hang thyself in thine own heir-apparent garters! If I be ta'en, I'll peach for this. An I have not ballads made on you all, and sung to filthy tunes, let a cup of sack be my poison.

(*Enter* GADSHILL)

GADSHILL. Stand!

FALSTAFF. So I do, against my will.

POINTZ. O, 'tis our setter: I know his voice.

(*Coming forward with* BARDOLPH *and* PETO)

BARDOLPH. What news?

GADSHILL. On with your vizards: there's money of the king's coming down the hill: 'tis going to the king's exchequer.

FALSTAFF. Ye lie, ye rogue; 'tis going to the king's tavern.

GADSHILL. There's enough to make us all.

FALSTAFF. To be hang'd.

HENRY. Sir, you four shall front them in the narrow lane; Ned Pointz and I will walk lower: if they escape from your encounter, then they light on us.

PETO. How many be there of them?

GADSHILL. Some eight or ten.

FALSTAFF. Zounds, will they not rob us?

HENRY. What, a coward, Sir John Paunch?

FALSTAFF. Well, I am not John of Gaunt, your grandfather; but yet no coward, Hal.

HENRY. Well, we leave that to the proof.

POINTZ. Sirrah, Jack, thy horse stands behind the hedge: when thou need'st him, there shalt thou find him. Farewell, and stand fast.

FALSTAFF. Now cannot I strike him, if I should be hanged!

HENRY. (*Aside to* POINTZ)   Ned, where are our disguises?

POINTZ. (*Aside to* PRINCE HENRY)   Here, hard by. Stand close.

(*Exeunt* PRINCE HENRY *and* POINTZ)

FALSTAFF.   Now, my masters, happy man be his dole, say I:
every man to his business.

FIRST TRAVELLER.   Come neighbour; the boy shall lead our
horses down the hill; we'll walk a-foot awhile, and
ease our legs.

THIEVES.   Stand!

TRAVELLERS.   Jesus bless us!

FALSTAFF.   Strike; down with them, cut the villains' throats: ah,
whoreson caterpillars! bacon-fed knaves! they hate us
youth;—down with them; fleece them.

TRAVELLERS.   O, we are undone, both we and ours for ever.

FALSTAFF.   Hang ye, gorbellied knaves, are ye undone? No, ye
fat chuffs; I would your store were here! On, bacons,
on! What, ye knaves! young men must live.

(*They rob and bind them. Exeunt.*
Enter PRINCE HENRY *and* POINTZ *in disguise*)

HENRY.   The thieves have bound the true men. Now could thou
and I rob the thieves, and go merrily to London, it
would be laughter for a month, and a good jest for ever.

POINTZ.   Stand close: I hear them coming.   (*They retire*)

(*Enter* THIEVES *again*)

FALSTAFF.   Come, my masters, let us share, and then to horse
before day. An the Prince and Pointz be not two arrant
cowards, there's no equity stirring: there's no more
valour in that Pointz than in a wild duck.

HENRY.   Your money!

POINTZ.   Villains!

(*As the* THIEVES *are sharing the* PRINCE *and* POINTZ *set upon them.*
They all run away, leaving the booty behind them and FALSTAFF, *after*
a blow or two, runs away too)

HENRY. Got with much ease. Now merrily to horse:
The thieves are scattered, and possess'd with fear
So strongly that they dare not meet each other;
Each takes his fellow for an officer.
Away, good Ned. Falstaff sweats to death,
And lards the lean earth as he walks along;
Were't not for laughing, I should pity him.

POINTZ. How the rogue roar'd!   (*Exeunt*)

SCENE III

Eastcheap. The Boar's Head Tavern

(*Enter* PRINCE HENRY)

HENRY. Ned, prithee, come out of that fat room, and lend me
thy hand to laugh a little. (*Enter* POINTZ) Falstaff and
the rest of the thieves are at the door: shall we be
merry?

POINTZ. As merry as crickets, my lad. But hark ye...

HENRY. Call in ribs, call in tallow...

(*Enter* FALSTAFF, GADSHILL, BARDOLPH, PETO *and tapster with wine*)

POINTZ. Welcome Jack: where hast thou been?

FALSTAFF. Plague of all cowards, I say, and a vengeance too!
Marry, and amen!—Give me a cup of sack, boy. A
plague of all cowards!—Give me a cup of sack, rogue.
—Is there no virtue extant? (*Drinks*) You rogue, here's
lime in this sack too: there is nothing but roguery to be
found in villainous man: yet a coward is worse than a
cup of sack with lime in it,—a villainous coward. Go
thy ways, Old Jack.... There lives not three good men
unhang'd in England; and one of them is fat, and
grows old: God help the while! A bad world I say.
A plague of all cowards! I say still.

HENRY. How now, wool-sack! what mutter you?

FALSTAFF.   A king's son! If I do not beat thee out of thy kingdom
with a dagger of lath, and drive all thy subjects afore
thee like a flock of wild geese, I'll never wear hair on
my face more. You Prince of Wales!

HENRY.   Why, you whoreson round man, what's the matter?

FALSTAFF.   Are you not a coward? answer me to that:—and
Pointz there?

POINTZ.   Zounds, ye fat paunch, an ye call me coward, by the
Lord, I'll stab thee.

FALSTAFF.   I call thee coward! I'll see thee damn'd ere I call thee
coward: but I would give a thousand pound, I could
run as fast as thou canst. You care not who sees your
back: call you that backing of your friends? A plague
upon such backing. Give me them that will face me.
Give me a cup of sack. I am a rogue if I drunk today!

HENRY.   O villain! thy lips are scarce wiped since thou drunkst
last.

FALSTAFF.   All's one for that. A plague of all cowards! Still say
I.   (*Drinks*)

HENRY.   What's the matter?

FALSTAFF.   What's the matter! there be four of us here have ta'en
a thousand pound this day morning.

HENRY.   Where is it, Jack? Where is it?

FALSTAFF.   Where is it! taken from us it is: a hundred upon poor
four of us.

HENRY.   What, a hundred, man?

FALSTAFF.   I am a rogue, if I were not at half-sword with a dozen
of them two hours together. I have scaped by miracle.
I am eight times thrust through the doublet, four
through the hose: my buckler cut through and through;
my sword hacked like a hand-saw: I never dealt better
since I was a man: all would not do.

HENRY.   Speak, sirs, how was it?

GADSHILL. We four set upon some dozen,—

FALSTAFF. Sixteen at least, my lord.

GADSHILL. And bound them.

PETO. No, no, they were not bound.

FALSTAFF. You rogue, they were bound, every man of them, or
I am a jew else, an Ebrew jew.

GADSHILL. As we were sharing, some six or seven fresh men set
upon us,—

FALSTAFF. And unbound the rest, and then come in the other.

HENRY. What, fought you with them all?

FALSTAFF. All! I know not what you call all: but if I fought not
with fifty of them, I am a bunch of radish: if there were
not two or three and fifty upon poor old Jack, then I
am no two-legged creature.

HENRY. Pray God you have not murder'd some of them.

FALSTAFF. Nay, that's past praying for: I have peppered two of
them; two I am sure have paid,—two rogues in
buckram suits. I tell thee what, Hal,—if I tell thee a lie,
spit in my face, call me horse. Thou knowest my old
ward,—here I lay, and thus I bore my point. Four
rogues in buckram let drive at me,—

HENRY. What, four? thou saidst but two even now.

FALSTAFF. Four, Hal, I told thee four.

POINTZ. Ay, ay, he said four.

FALSTAFF. These four came all afront, and mainly thrust at me. I
made no more ado, but took all their seven points in
my target, thus.

HENRY. Seven? why there were but four even now.

FALSTAFF. In buckram?

POINTZ. Ay, four in buckram suits.

FALSTAFF. Seven, by these hilts, or I am a villain else.

HENRY. Prithee, let him alone; we shall have more anon.

FALSTAFF. Dost thou hear me, Hal?

HENRY. Ay, and mark thee too, Jack.

FALSTAFF. Do so, for it is worth the listening to. These nine in buckram that I told thee of,—

HENRY. So, two more already.

FALSTAFF. Their points being broken...began to give ground: but I followed me close, came in foot and hand; and with a thought seven of the eleven I paid.

HENRY. O monstrous! eleven buckram men grown out of two.

FALSTAFF. But, as the devil would have it, three misbegotten knaves in Kendal green came at my back and let drive at me;—for it was so dark, Hal, that thou couldst not see thy hand.

HENRY. These lies are like their father that begets them,— gross as a mountain, open, palpable. Why, thou clay-brained guts, thou nott-pated fool, thou whoreson, obscene, greasy tallow-keech,—

FALSTAFF. Why, art thou mad? art thou mad? is not the truth the truth?

HENRY. Why, how couldst thou know these men in Kendal green, when it was so dark thou couldst not see thy hand? come, tell us your reason: what say'st thou to this?

POINTZ. Come, your reason, Jack, your reason.

FALSTAFF. What, upon compulsion? Zounds, and I were at the strappado, or all the racks in the world, I would not tell you on compulsion. Give you a reason on compulsion! if reasons were as plentiful as blackberries, I would give no man a reason upon compulsion, I.

HENRY. I'll be no longer guilty of this sin; this sanguine coward, this bed-presser, this horse-back breaker, this huge hill of flesh,—

FALSTAFF. Away you starveling, you eel-skin, you dried neat's tongue, you stock-fish,—O for a breath to utter what is like thee!—you tailor's yard, you sheath, you bow-case, you vile....

HENRY.	Well, breathe awhile, and then to it again: and when thou hast tired thyself in base comparisons hear me speak but this.

POINTZ.	Mark, Jack.

HENRY.	We two saw you four set on four and bound them, and were masters of their wealth,—Mark now, how a plain tale shall put you down.—Then did we two set on you four; and, with a word, outfaced you from your prize, and have it; yea, and can show it here in the house;—and Falstaff, you carried your guts away as nimbly, with as quick dexterity, and roar'd for mercy, and still ran and roar'd, as ever I heard bull-calf. What a slave art thou, to hack thy sword as thou hast done, and then say it was in fight! What trick, what device, what starting-hole, canst thou now find out to hide thee from this open and apparent shame?

POINTZ.	Come, let's hear, Jack: what trick hast thou now?

FALSTAFF.	By the Lord, I knew ye as well as he that made ye. Why, hear you my masters; was it for me to kill the heir-apparent? should I turn upon the true prince? why, thou knowest I am as valiant as Hercules: but beware instinct; the lion will not touch the true prince. Instinct is a great matter; I was a coward on instinct. I shall think the better of myself and thee during my life; I for a valiant lion and thou for a true prince. But, by the Lord, lads, I am glad you have the money. Hostess, clap-to the doors:—watch tonight, pray tomorrow.—Gallant lads, boys, hearts of gold, all the titles of good fellowship come to you! What, shall we be merry? shall we have a play extempore?

HENRY.	Content: and the argument shall be thy running away.

FALSTAFF.   Ah, no more of that, Hal, an thou lovest me!

(*Enter* HOSTESS)

HOSTESS.   O Jesus, my lord the prince—

HENRY.   How now, my lady the hostess! What sayst thou to me?

HOSTESS.   Marry my lord, there is a nobleman of the court at door would speak with you: he says he comes from your father.

HENRY.   Get him as much as will make him a royal man, and send him back again to my mother.

FALSTAFF.   What manner of man is he?

HOSTESS.   An old man.

FALSTAFF.   What doth gravity out of his bed at midnight?— Shall I give him an answer?

HENRY.   Prithee do, Jack.

FALSTAFF.   Faith, and I'll send him packing.

(*Exit* FALSTAFF)

HENRY.   Now sirs, by'r lady, you fought fair; so did you, Peto, so did you, Bardolph; you are lions, too, you ran away upon instinct, you will not touch the true prince; no, fie!

BARDOLPH.   Faith, I ran when I saw the others run.

HENRY.   Faith, tell me in earnest, how came Falstaff's sword so hack'd?

PETO.   Why, he hack'd it with his dagger, and said he would swear truth out of England, but he would make you believe it was done in fight: and persuaded us to do the like.

BARDOLPH.   Yea, and to tickle our noses with spear-grass to make them bleed; and to beslubber our garments with it, and swear it was the blood of true men. I did that I did not these seven year before,—I blush'd to hear his monstrous devices.

HENRY.   Here comes lean Jack, here comes bare-bone.

(*Enter* FALSTAFF)

How now, my sweet creature of bombast! how long is't ago, Jack, since thou saw'st thine own knee?

FALSTAFF. My own knee! When I was about thy years, Hal. A plague of sighing and grief! it blows a man up like a bladder. There's villainous news abroad; here was Sir John Bracy from your father; you must to the Court in the morning. That same mad fellow of the North, Percy, was there too, and he of Wales... and one Mordlake, and a thousand blue-caps more: Worcester is stolen away tonight; thy father's beard is turn'd white with the news: you may buy land now as cheap as stinking mackerel. *(A knocking heard)*

(Exeunt HOSTESS *and* BARDOLPH. *Re-enter* BARDOLPH *running)*

BARDOLPH. O my lord, my lord! the sheriff with a most monstrous watch is at the door.

(Enter HOSTESS *hastily)*

HOSTESS. O Jesu, my lord, my lord,—

HENRY. Heigh, heigh! the devil rides upon a fiddlestick: what's the matter?

HOSTESS. The sheriff and the watch are at the door: they are come to search the house. Shall I let them in?

FALSTAFF. Dost thou hear, Hal? never call a piece of true gold a counterfeit: thou art essentially mad, without seeming so.

HENRY. And thou a natural coward, without instinct. Go hide behind the arras: the rest walk up above. Now my masters, for a true face and good conscience.

FALSTAFF. Both of which I have had: but the date is out, and therefore I'll hide me.

HENRY. Call in the sheriff.

(Exeunt all except the PRINCE *and* POINTZ. *Enter* SHERIFF *and* CARRIER)
Now master Sheriff, what's your will with me?

SHERIFF. First, pardon me, my lord. A hue and cry
Hath followed certain men unto this house.

HENRY. What men?

SHERIFF. One of them is well-known, my gracious lord, a gross fat man.

CARRIER. As fat as butter.

HENRY. The man, I do assure you, is not here;
And, sheriff, I will engage my word to thee,
That I will, by tomorrow dinner-time,
Send him to answer thee, or any man,
For anything he shall be charg'd withal:
And so, let me entreat you leave the house.

SHERIFF. I will, my lord. There are two gentlemen
Have in this robbery lost three hundred marks.

HENRY. It may be so: if he have robb'd these men,
He shall be answerable; and so, farewell.

SHERIFF. Goodnight, my noble lord.

HENRY. I think it is good morrow, is it not?

SHERIFF. Indeed, my lord, I think it be two o'clock.
(*Exeunt* SHERIFF *and* CARRIER)

HENRY. This oily rascal is known as well as Paul's.
Go call him forth.

POINTZ. Falstaff!—fast asleep behind the arras, and snorting like
a horse.

HENRY. Hark, how hard he fetches breath. Search his pockets.
(*He searcheth his pockets and findeth certain papers*)
What hast thou found?

POINTZ. Nothing but papers, my lord.

HENRY. Let's see what they be: read them.

POINTZ. (*reads*)
Item, A capon	2s.	2d.
Item, Sauce		4d.
Item, sack, two gallons	5s.	8d.
Item, Anchovies and sack		
after supper	2s.	6d.
Item, Bread		ob.[1]

---

[1] An abbreviation for 'obulus', a halfpenny.

HENRY. O monstrous! but one half-pennyworth of bread to this intolerable deal of sack!—What there is else, keep close; we'll read it at more advantage: there let him sleep till day. I'll to the court in the morning. We must all to the wars, and thy place shall be honourable. I'll procure this fat rogue a charge of foot; and I know his death will be a march of twelve-score. The money shall be paid back again with advantage. Be with me betimes in the morning; and so, good-morrow, Pointz.

POINTZ. Good morrow, good my lord.
(*Exeunt*)

END OF EXTRACT

APPENDIX I

A Programme about Childhood

NOTE

This may be done with four voices, or more, players for chime bars or percussion if you have them, singers and a pianist to play the pieces of music.

*　　　*　　　*

The sources of the material are: (1) Ecclesiastes xi. 9. (2) Traditional lullaby from *A Book of Nursery Songs*, S. Baring Gould, Methuen, 1895. (3) *Animula*, T. S. Eliot. (4) *Portrait of the Artist as a Young Man*, James Joyce, first page. (5) *The Manchester Guardian*. (6) *Children's Games* by the present author. (7) See list of folksong gramophone records, *Folksongs for Use in Schools*, No. 3, Novello. (8) *Huckleberry Finn*, Mark Twain. (9) *I saw Esau*, I. and P. Opie. (10) Traditional. (11) *The Retreat*, Henry Vaughan. (12) *Portrait of the Artist as a Young Dog*, Dylan Thomas. (13) *Landscapes*, T. S. Eliot. (14), (15) Traditional: from *The Poet's Tongue*. (16) *Children's Games*. (17) Baring Gould, *op. cit.* (18) Traditional: from *The Poet's Tongue*. (19) *Oxford Dictionary of Nursery Rhymes*. (20) *Fern Hill*, Dylan Thomas. (21) *Oxford Dictionary of Nursery Rhymes*. (22) Traditional: from *The Poet's Tongue*. (23) Baring Gould, *op. cit.* (24) Ecclesiastes xi. 9.

*　　　*　　　*

I have in this programme deliberately avoided being explicit— by way of having an 'announcer' or introductory talk. The impulse to have everything explained (sometimes explained away) belongs to the mass media and the steady movement towards a lower common denominator in order to hold a mass audience. Thus we become less used to taking the implicit meaning of the work of art, its direct impact, its particular modes of presentation. The folk-mind would not need to be told that the Revesby

Appendix I

Plough Play is about the death of the old year and the birth of the new, about the magical way in which the old and dead becomes in the spring the young, leaping and fertile new; or how the Fool is the intractable power of life which always re-asserts itself. John Barleycorn and the Green Man they recognised as the same kind of mad poetic figure as Mak the Sheep-stealer in *The Wakefield Shepherds' Play* and Autolycus in *A Winter's Tale*. We find it hard to understand these plays. Children and the unsophisticated primitive can accept their meaning implicitly.

This collection is intended to make a semi-dramatic programme of extracts from poems and passages evocative of states of childhood. It is an attempt to show the way the child makes its way 'over the one-strand river' into adulthood by the very rhythm, nonsense, and wisdom of the traditional lore of childhood. Some passages suggest the terrors inevitably met with, others the inevitable clouding of innocence, others the great strength of the dream vision—*Nottamun Town* seems to contain a great and mysterious wisdom, but hardly any explicit meaning. The programme could be extended by drawing on, for example, William Blake, Lewis Carroll, Walter de la Mare, Wordsworth, Yeats (*A Prayer for my Daughter*), Edward Thomas (*Old Man*), D. H. Lawrence (*As a drenched drowned bee*), Edmund Gosse (*Father and Son*), Alison Uttley (from *A Country Child*), the Opies' books, Mark Twain, and the poetry of James Reeves.

The material chosen is delicate but not whimsical or fey, I trust. Yet there is no need for us to assume that under their sophistication our children are anything other than little children. One may find, now and then, a crowd of callow boys in assembly as moved as one is oneself by a small girl reading, say, the account of Abraham and Isaac or the passage from *Ruth*, Chapter One. Such experiences confirm one's faith in children's essential tenderness, and its persistence even under the affected cynicism imposed on them sometimes by an industrial-commercial culture.

158

1 Rejoice, O young man, in thy youth; and let thy heart cheer thee in the days of thy youth, and walk in the ways of thine heart, and in the sight of thine eyes:

2 (*Sung to chime bars*)
>   No silk was ever spun so fine
>   As is the hair of baby mine.
>   My baby smells more sweet to me
>   Than smells in spring the elder tree.
>   And it's O sweet sweet and a lullaby.

3   Issues from the hand of God, the simple soul
>   To a flat world of changing lights and noise,
>   To light, dark, dry or damp, chilly or warm;
>   Moving between the legs of tables and of chairs,
>   Rising or falling, grasping at kisses and toys,
>   Advancing boldly, sudden to take alarm,
>   Retreating to the corner of arm or knee,
>   Eager to be reassured, taking pleasure
>   In the fragrant brilliance of the Christmas tree,—
>   Pleasures in the wind, the sunlight and the sea. . . .

4 Once upon a time and a very good time it was there was a moocow coming down along the road and this moocow that was coming down the road met a little boy called a baby cuckoo.... His father told him that story: his father looked at him through a glass: he had a hairy face.... His mother had a nicer smell than his father.

5 It was stated at East London Juvenile Court yesterday that when police visited a flat at Stepney, they found that the food cupboard contained only two loaves of bread, two bottles of

milk, and some porridge oats. The nine children of the family, who were alone in the flat...aged from one to eleven...were sitting watching television.

6   *A:* Where are your manners?
      B: In my Shoe!
      C: Who do you care for?
     A,B,C: NOT FOR YOU!

7 *(Sung)*   Dance to your daddy,
       My little laddie.
       Dance to your daddy,
       My little man.
       Thou shalt have a fish,
       Thou shalt have a fin,
       Thou shalt have a haddock,
       When the boat comes in.
       Thou shalt have a codling
       Boilèd in a pan.
       Dance to your daddy,
       My little man.

       When thou art a man,
       And fit to take a wife,
       Thou shalt wed a maiden
       Love her all your life.
       She shall be your lassie—
       Thou shalt be her man,
       Dance to your daddy
       My little man.

8 All of a sudden, bang! bang! bang! goes three or four guns—the men had slipped around through the woods and come in from behind with their horses! The boys jumped for the river—both of them hurt—and as they swum down the current the men run along the bank shooting at them and singing out, 'Kill them! Kill them!' It made me so sick I 'most fell out of the tree. I was

powerful glad to get away from the feuds...there weren't no
home like a raft, after all. Other places seem so cramped up and
smothery, but a raft don't. You feel mighty free and easy and
comfortable on a raft.

9   I had a black man, he was double-jointed
        I kissed him and made him disappointed.
        All right Hilda, I'll tell your mother,
        Kissing the black man round the corner.
        How many kisses did he give you?
        One, two, three
                (*carry on counting to twenty* sotto voce).

10 (*Sung to chime bars*)
           The hart he loves the high wood
           The hare she loves the hill;
           The knight he loves his bright sword,
           The lady loves her will.

11   Happy those early days, when I
        Shin'd in my Angel infancy....
        Before I taught my tongue to wound
        My conscience with a sinful sound,
        Or had the black art to dispense
        A several sin to every sense,
        But felt through all this fleshly dress
        Bright shoots of everlastingness....
        Oh how I long to travel back
        And tread again that ancient track....
        But oh! my soul with too much stay
        Is drunk, and staggers in the way....

12 I stole twelve books in three visits to the library and threw
them away in the park....I beat a dog with a stick so that it
rolled over and licked my hand afterwards....I saw Billy Jones
beat a pigeon to death with a fire shovel....

13   Children's voices in the orchard
Between the blossom and the fruit-time:
Golden head, crimson head,
Between the green tip and the root.
Black wing, brown wing, hover over;
Twenty years and the spring is over;
Today grieves, tomorrow grieves,
Cover me over, light in leaves;
Golden head, black wing,
Cling, swing,
Spring, sing,
Swing up into the apple tree.

14   *Here perhaps some music without words, e.g., a piano piece from*
From an Overgrown Path *by Janáček* (Goodnight *or* In Tears)*; or
from Schumann's* Scenes of Childhood.

15   (*Sung to chime bars*)

      A: This is the key of the kingdom;
In that kingdom is a city,
In that city is a town,
In that town there is a street,
In that street there winds a lane,
In that lane there is a yard,
In that yard there is a house,
In that house there is a room,
In that room there is a bed,
On that bed there is a basket,
    A basket of flowers.

      B: Flowers in the basket,
Basket on the bed,
Bed in the chamber,
Chamber in the house,
House in the weedy yard,

Yard in the winding lane,
Lane in the broad street,
Street in the high town,
Town in the city,
City in the kingdom:
    This is the key of the kingdom.

16 *A:*   Sunday, take care of Monday,
       Monday, take care of Tuesday,
       Tuesday, take care of Wednesday,
       Wednesday, take care of Thursday,
       Thursday, take care of Friday,
       Friday, take care of Saturday,
       Take care the Old Witch does not catch you, and
       I'll bring you something nice.

   B: Sunday, your mother sent me for your best bonnet,
       She wants to get one like it for Monday.
       It's up in the top drawer, fetch it quick.

   A: Where's Saturday?
   B and C: The Old Witch has got her!
   A and D: And Monday and Tuesday and Wednesday and
               Thursday and Friday....

   A: Have you seen my children?
   B: They are in bed.
   A: Can't I go up and see them?
   B: Your shoes are too dirty.
   A: Can't I take them off?
   B: Your stockings are too dirty.
   A: Can't I take them off?
   B: Your feet are too dirty.
   A: Can't I cut them off?
   B: The blood will run all over the floor.

A: Can't I wrap them up in a blanket?
B: The fleas would hop out.
A: Can't I wrap them up in a sheet?
B: The sheet is too white.
A: Can't I ride in a carriage?
B: You'd break the stairs down.
All: BURN THE OLD WITCH!

17   (*Sung to chime bars*)

My mother said that
I never should
Play with the gipsies
In the wood
If I did
She would say
Naughty girl
To disobey:
Your hair won't curl
Your shoes shan't shine:
You naughty girl
You shan't be mine.
My father said
That if I did
He'd bang my head
With a teapot lid.
The wood was dark
The grass was green,
In came Sally
With a tambourine.
Alapaca frock
New scarf shawl
White straw bonnet
And a pink parasol.
I went to the river
No ship to get across,

I paid ten shillings
For an old blind horse.
I up on his back
And off in a crack:
Sally tell my mother
I shall never come back.

On Saturday night be all my care
To powder my locks and curl my hair:
And Sunday morning my love will bring
To marry me fair with a golden ring.

18   There was a man of double-deed
Sowed his garden full of seed.
When the seed began to grow,
'Twas like a garden full of snow;
When the snow began to melt,
'Twas like a ship without a belt;
When the ship began to sail,
'Twas like a bird without a tail;
When the bird began to fly,
'Twas like an eagle in the sky;
And when the sky began to roar,
'Twas like a lion at the door;
When the door began to crack,
'Twas like a stick across my back;
When my back began to smart,
'Twas like a penknife in my heart;
When my heart began to bleed,
'Twas death and death and death indeed.

19   Birds of a feather flock together,
And so will pigs and swine
Rats and mice will have their choice,
And so will I have mine.

20 Nothing I cared in the lamb white days, that time would
                                                    take me
    Up to the swallow-thronged loft by the shadow of my
                                                        hand,
       In the moon that is always rising,
          Nor that riding to sleep
       I should hear him fly from the childless land.
    Oh as I was young and easy in the mercy of his means,
       Time held me green and dying
    Though I sang in my chains like the sea.

21 (*Sung to chime bars*)
                    I went to Noke
                    But nobody spoke
                    I went to Thame
                    It was just the same
                    Burford and Brill
                    Silent and still
                    Then I went to Beckley
                    And they spoke directly.

22 (*To a drum*)
    In Nottamun Town not a soul would look up,
    Not a soul would look up, not a soul would look down,
    Not a soul would look up, not a soul would look down,
    To tell me the way to Nottamun Town.

    I rode a big horse that was called a grey mare,
    Grey mane and tail, grey stripes down his back,
    Grey mane and tail, grey stripes down his back,
    There wasn't a hair on him but what was called black.

    She stood so still, she threw me to the dirt,
    She tore my hide and bruised my shirt;
    From stirrup to stirrup I mounted again
    And on my ten toes I rode over the plain.

Met the King and the Queen and a company of men
A-walking behind a-riding before,
A stark-naked drummer came walking along
With his hands in his bosom a-beating his drum.

Sat down on a hot and cold frozen stone,
Ten thousand stood round me yet I was alone.
Took my heart in my hand to keep my head warm.
Ten thousand got drowned that never were born.

23 Grey goose and gander, waft your wings together,
And carry the good King's daughter over the one-strand river.

24 Therefore remove sorrow from thy heart, and put away
evil from thy flesh, for childhood and youth are vanity.

THE END

APPENDIX II

THE TRADITIONAL PUPPET PLAY OF
Punch and Judy

FOREWORD

Piccini of Drury Lane is described in the Preface to the 1870
version of *The Tragical Comedy or Comical Tragedy of Punch and
Judy* as having 'perambulated town and country for the last forty
or fifty years'.

'He was an Italian; a little thick-set man, with a red humorous-
looking countenance. He had lost one eye, but the other made up
for the absence of its fellow by a shrewdness of expression
sufficient for both.[1] He always wore an oil-skin hat and a rough
greatcoat. At his back he carried a deal box containing the
dramatis personae of his little theatre; and in his hand the trumpet,
at whose glad summons, hundreds of merry laughter-loving
faces flocked round him with gaping mouths and anxious looks,
all eager to renew their acquaintance with their old friend and
favourite, Mr. Punch. The theatre itself was carried by a tall man,
who seemed a sort of sleeping partner in the concern, a mere
"dumb waiter" on the other's operations.'[2]

Payne Collyer corrected Piccini's script, and added songs from
a manuscript of 1796. I have re-written some of the songs in the
same rhythm and mode; some obviously derive from *The
Beggar's Opera*, which was first acted in 1728, and other popular
operas. Of course the songs in *Punch* change with popular
taste—but modern dance-song modes always seem to me un-
suitable.

According to Payne Collyer Piccini's puppets were more
expertly carved from wood than those of other performers, and
he brought them over from Italy, complaining that he could not
find any workmen in England capable of repairing a broken

---

[1] The description makes me feel that it was possibly to Piccini that Dickens owes his
inspiration for Sleary in *Hard Times*. The account comes from *Literary Speculum* earlier in
the nineteenth century.

[2] In the trade he is called 'the Bottler' and is not dumb, but acts as a kind of 'feed'.

figure or replacing a stolen one. (My drawings are from Cruik-shank's engravings of Piccini's puppets.)

Piccini's exhibition was originally all given in Italian; only later did he 'translate' it into English. Other native Punch and Judy men, apparently recognising Punch as of Italian origin, generally at that time imitated an 'outlandish origin' for Punch. It was Payne Collyer who divided Mr Punch into acts and scenes; he should be played straight through, like Shakespeare, but I have kept acts and scenes to give young performers a breather.

THE CHARACTERS

PUNCH

LORD CHIEF JUSTICE OF ENGLAND

SCARAMOUCHE   JACK KETCH   THE CHILD

THE DEVIL   BLIND MAN   TOBY

DOCTOR   HECTOR   BLACK SERVANT

JUDY   CONSTABLE   POLLY

POLICE OFFICER   THE BOTTLER

GHOST

The BOTTLER *is the gentleman who stands in front of the theatre and talks to the audience and to* PUNCH. *Once the play begins he goes away. But he is there to warm things up and to 'feed'* MR PUNCH *with some good lines for him to make jokes with. The curtains are drawn when the* BOTTLER *blows a trumpet and begins:*

BOTTLER.   Come along now everyone: we all know what we're here for, don't we? (*no answer*) Don't we?

SOMEONE IN AUDIENCE.   Ye-es.

BOTTLER.   And what are we here for?

AUDIENCE. (*feebly*)   Mister Punch.

BOTTLER.   Now, boys and girls, my friend Mr Punch is having a nap in there. Now I'm not going to wake him up if all you can say is (*whispers and looks feeble*) 'Mister Punch'. You don't sound as if you cared a halfpenny whether Mr Punch appeared or not. You don't care do you?

HALF THE AUDIENCE.   } (*weakly*) { Yes.

OTHER HALF OF THE AUDIENCE. }   { No.

173

Punch and Judy

BOTTLER. You don't care if we go away and leave Mr Punch asleep, do you?

AUDIENCE. (*alarmed*) { Yes. / No. }

BOTTLER. Shall we pack up then?

AUDIENCE. (*really worried*) No!

BOTTLER. All right then. But you've got to do better this time. We'll have to wake old Punch up. Right?

AUDIENCE. Yes.

BOTTLER. Right then. Now when I count three you shout, 'We want Punch'. Ready. One! Two! Three!

AUDIENCE. (*feebly*) We want Punch!

BOTTLER. No! No good at all. (*Mimics them in a thin voice*) 'We want Punch!' You sound as if you're asking your Mum for a dose of paragoric embrocation for superfluous tonsils. Come on now, take a deep breath, hold your ears back, take out your gob stoppers, and ONE, TWO, THREE!

AUDIENCE. WE WANT PUNCH!

PUNCH. (*off, which means he is still inside the booth behind the curtains*) Oi de doi de diddleydoi![1]

BOTTLER. Mr Punch!

PUNCH. 'Ullo. What you waking me up for?

BOTTLER.[2] What are we waking you up for? We want a song.

PUNCH. You want a song? What for?

BOTTLER. What for? Well...er.... What do we want a song for, boys and girls? I know, to cheer us up.

PUNCH. I can't cheer anybody up. I'm miserable.

---

[1] When Punch makes this noise it is something between a 'Hurray!' and 'He! He! He! He!'—a cry of exultation. It is made with the instrument the Punch and Judy man uses in his mouth to talk with, pressed against his soft palate. It is called in the trade a SWAZZLE: properly it is called, in French, *un sifflet practique*.

[2] The Bottler repeats everything Punch says for a while so the audience get used to Punch's very squeaky way of talking.

174

BOTTLER. You're miserable! What's the matter, Mr Punch?

PUNCH. Working too hard.

BOTTLER. Working too hard? What have you been doing?

PUNCH. Dreaming.

BOTTLER. Dreaming? That's not hard work.

PUNCH. Oh, yes it is! I was dreaming I was working.

BOTTLER. Come now, Mr Punch, dreaming you were working doesn't make you tired.

PUNCH. Oh, yes it does!

BOTTLER. Oh, no it doesn't. Does it boys and girls?

AUDIENCE. { Yes.
{ No.

BOTTLER. Well, what are you going to do now, Mr Punch?

PUNCH. Wake up and have a nice rest.

BOTTLER. So if we hadn't woken you up, Mr Punch, you'd have been worked to a shadow: so come out now and say 'hallo' to the audience just to show how grateful you are.

PUNCH. Shan't!

BOTTLER. Mr Punch, I'm afraid you're incorrigible.

PUNCH. What's that?

BOTTLER. Well it's, er, it means, er...tell him boys and girls.

AUDIENCE. Don't know.

BOTTLER. Well it means...you're...incorrigible. And I'm afraid, boys and girls, it means we shan't be having a show at all, so you can all be taking your money back that we collected to buy Mr Punch a new stick with, with bells on the end and coloured ribbons....

PUNCH. (*comes out and knocks* BOTTLER's *top hat off*) Oi de doi de doi! That's the way to do it!

BOTTLER. So there you are. What were you waiting for?

PUNCH. I was waiting till you'd said you'd buy me a new stick.

BOTTLER. Did I say I'd buy him a new stick?

AUDIENCE. Yes!

BOTTLER. All right. Well, you'd better sing for it, Punch; no
new stick without some *mu-sic*. Ha! Ha! ha!

(*Picks up his hat and puts it on. Punch knocks it off again*)

PUNCH. That's the way to do it!

BOTTLER. Well, I'm not going to have any more to do with you.
I shall leave you with these boys and girls and I hope
they give you your rightful deserts.

PUNCH. Trifle dessert? Come on then, boys and girls.

BOTTLER. I did not say trifle dessert. I said the bird, the cold
shoulder, the raspberry, *rightful deserts.*

PUNCH. Cold shoulder, cold bird, raspberry trifle dessert. Oi de
doi, (*sings*) we'll have the lot for dinner tonight.

BOTTLER. Well there he is, boys and girls; I can't do anything
with him, perhaps you can.

(*Exit* BOTTLER)

ACT I

(PUNCH *dances about the stage*)

PUNCH. Hurrah for old Punch, he's the prince of all fellows,
His costume magnificent scarlets and yellows,
His nose like a fire and his cheeks like a bellows.
He don't care for no one, he don't want no friends,
He's fierce and he's fat, he's a rogue and a rover,
He don't care for nothing, but living on clover,
(*Spoken*) Until he drops dead and the whole thing is over—
(*Mock sad*) And there Punch's comedy ends.

(*He goes on dancing, calling 'Judy my dear! Judy!'*)

(*Enter* DOG TOBY)

PUNCH. Hallo Toby! Who called you? How do you do,
Mr Toby? Hope you're very well, Mr Toby.

Punch and Judy

TOBY. Wow! Wow! Wow!

PUNCH. And how's my good friend your master, Mr Toby? How's Mr Scaramouche?

TOBY. Wow! Wow! Wow!

PUNCH. I'm glad to hear it. Poor Toby! What a nice good-tempered dog it is! No wonder his master is so fond of him.

TOBY. (*snarls*) Brr! Brr!

PUNCH. What! Toby! Not know your kind friend Punch?

TOBY. (*snarls again*) Arr! Arr! Arr!

PUNCH. There, there, now now Toby. (*putting out his hand cautiously and trying to coax the dog*) Toby, you're a nasty cross dog. I'll send you to the nice kind vet to put you to sleep: get off home, dog. (*strikes at* TOBY)

TOBY. Bow! wow! wow! (*Seizes* PUNCH *by the nose*)

PUNCH. Oh! Oh dear! My nose! My poor nose! My beautiful nose! Get away! Get away—you nasty dog. I'll tell your master! Help! Judy! Judy!

(PUNCH *shakes his nose, but he can't shake off* TOBY, *who clings on to his nose as* PUNCH *goes backwards round the stage.* PUNCH *continues to shout 'Judy! Judy, my dear!' until* TOBY *lets go and goes out*)

PUNCH. (*alone, rubs his nose with both hands*) Oh, my nose! My pretty little nose. Judy, that dog's damaged the best part of me! Judy! The nasty brute: I'll tell its master! Mr Scaramouche! Look what your nasty brute of a dog has done!

(*Enter* SCARAMOUCHE *who has a tall hat, a hook nose and a hump, like* PUNCH, *and a stick*)

SCARAMOUCHE. Hallo, Mr Punch. What have you been doing to my poor dog Toby?

(*Looking at* SCARAMOUCHE's *stick,* PUNCH *is scared and steals backwards till he disappears behind the curtains stage right—which means those on the right-hand side of the opening as you look out at the audience—and then puts his head round the outside of the front of the theatre)*

PUNCH. Ah, my good friend, how do you do? Nice to see you looking so well. *(Aside: that is, to the audience so* SCARAMOUCHE *can't hear)* I wish you were miles away with that great stick.

SCARAMOUCHE. You've been beating and aggravating my poor dog, Mr Punch.

PUNCH. He has been biting and aggravating my poor nose. What have you got there?

SCARAMOUCHE. Where?

PUNCH. In your hand.

SCARAMOUCHE. A fiddle. Can you play the fiddle?

PUNCH. *(Coming back on to the stage and going up to him)* I don't know: never tried. Let me see. You hold it like this? *(He takes the stick and moves slowly about to the tune of the 'Marseillaise')* And bring it round like this? *(He hits* SCARAMOUCHE *a slight blow on the hat as if by accident)*

SCARAMOUCHE. You play very well, Mr Punch. Now let me try. I will give you a little lesson on the fiddle. *(He hits* PUNCH*)*

PUNCH. Now let me try. *(He dances even faster, gets behind* SCARAMOUCHE *and gives him such a blow on the back of his head that he knocks it clean off his shoulders)*

Punch and Judy

PUNCH. How do you like that tune, eh? Sweet music or sour music or mustard music or vinegar music, eh? He! He! He! That's the way to do it. (*Throws away the stick inside the booth*) You'll never hear such another lovely tune as long as you live, me boy! (*Sings his first tune*) Judy, Judy, my dear! Judy, can't you answer a man, my dear?

JUDY. ('*Within*'—*that is, she is not yet seen on the stage*) What do you want, Mr Punch?

PUNCH. Come upstairs: I want you.

JUDY. Then want must be your mistress. I'm busy.

PUNCH. (*sings*)
Her tongue is polite and ever so civil
Her temper is sweet, her head cool and level
Yet sometimes I wish she would go to the Devil
Since that's all the answer I get!
And really I'm sorry to say, sirs,
That that is too often her way, sirs,
For this by and by she shall pay, sirs,
Oh wives are an obstinate set!

Judy my dear! (*Shouting*) Judy, my love, pretty Judy, come upstairs!

(*Enter* JUDY)

JUDY. Well, here I am! What d'you want, now I'm here?

PUNCH. (*aside*) What a pretty creature! Ain't she a beauty?

JUDY. What d'you want, I say?

PUNCH. A kiss! A luverly kiss!
(*He kisses her and she boxes his ears*)

JUDY. Take that then! How do you like my kisses? Like another?

PUNCH. No: one at a time, one at a time, my sweet pretty wife. (*Aside*) She's always so playful. (*To* JUDY) Where's the child? Fetch me the child, Judy, my dear.

(*Exit* JUDY)

HT

Punch and Judy

There's a whale of a wife for you! What a precious darling creature! She has gone to fetch our dear little baby!

(Enter JUDY *with the* CHILD*)*

JUDY. Here's the baby. Pretty dear! It knows its dad. Take the child.

PUNCH. *(holding out his hands)* Give it to me—pretty little thing! How it's like its dear sweet mummy!

JUDY. How clumsy you are! You're holding it upside down.

PUNCH. Oh no I'm not!

JUDY. Oh yes you are!

PUNCH. Oh no I'm not!

JUDY. Oh yes you are!

(They fight over the baby, and tug to and fro to get possession of it. Then they gradually change from tugging to rocking to and fro)

PUNCH. Hush-a-by, baby, on the tree top,
When the wind blows the cradle will drop.
*(*PUNCH *tugs baby away and tosses it in the air)*
'At's the way to do it. Whoops! Baby bunting.

JUDY. Punch! *(He catches it and rocks it again)*

PUNCH. Get away! I know how to nurse it as well as you do. Judy dear, go and peel the rhubarb for dinner. I'll mind the fruits of our awful union.

JUDY. *(aside)* Now children, I must go and cook the dinner. I don't trust that man with my dear little baby, so if he does anything horrid or cruel to it, you'll call me won't you?

AUDIENCE. Yes.

JUDY. That's right. Now you be patient, Punch. *(Exit)*

PUNCH. Oh go to sleep baby,
Your daddy is here;
Your mummy's gone crazy,
She's out on the beer.

Oh rest you my darling,
Mummy's coming home later—
With a voice like a starling
Or an old nutmeg grater.

Poor dear little thing. It can't get to sleep. By, by,
hush-a-by. All right, it shan't bother to go to sleep.
(*He dandles the baby on his knee*)

Dancy, baby, dancy
How it shall gallop and prancy!
Sit on my knee
And now kissy me
Dancy, baby, dancy.

(*But the* CHILD *begins to wail, possibly upset by the nearness of*
PUNCH's *enormous nose.* PUNCH *then sticks the* CHILD *against the side
of the stage, on the platform, and going himself to the opposite side, runs
up to it, clapping his hands and crying* 'Catchee, catchee, catchee!'
He then takes it up again and it begins to cry again)

What is the matter with it? Poor thing! It has got the
stomach ache, I dare say. (CHILD *cries*) Hush-a-by,
hush-a-by. (*sitting down and rolling the child on his knees*)
Naughty child! Judy! (*calling*) the child has got the
stomach ache. Phew! Nasty child! Judy! I say.
(CHILD *continues to cry*) Keep quiet, can't you? (*Hits
it a box on the ear*) Oh you filthy child! What have you
done? I won't keep such a nasty child. Hold your
tongue! (*Strikes the* CHILD's *head several times against
the side of the stage*) There! There! There! How do you
like that? I thought I'd stop your squalling. Get along
with you, nasty, naughty, crying child. (*Throws it
over the front of the stage among the audience who are
shouting* 'Judy!') He! He! He! Oi de diddley doi!
'At's the way to do it! Oi-de-doi-de-diddley-doi!
He! He!

(*Re-enter* JUDY)

7-2

JUDY.   Where is the child?

PUNCH.   Gone—gone to sleep. Oi de doi de doi!

AUDIENCE. (*variously*)   He's killed it. He's thrown it out, etc.

JUDY.   What have you done with my baby, I say?

PUNCH.   Gone to sleep, I say.

JUDY.   What have you done with it?

PUNCH. (*mock-incredulous*)   What have I *done* with it?

JUDY.   Yes. *Done* with it. I heard it crying just now. Where is it?

PUNCH.   How should I know?

AUDIENCE.   Oh!

JUDY.   I heard you make the pretty darling cry. What did he do, boys and girls?

AUDIENCE. (*variously describes what* PUNCH *did*)   Dropped it out, etc.

PUNCH.   I dropped it out of the window.

JUDY.   Oh you cruel horrid wretch, to drop the pretty baby out of the window. Oh! (*Cries, and wipes her eyes with the corner of her white apron*)   You wicked man! Oh!

PUNCH.   You shall have another one soon Judy my dear. Plenty more babies where that one came from. Common as flies. Oi de doi de doi.

JUDY.   I'll make you pay for this, depend upon it.
                                (*Exit in haste*)

PUNCH.   Off she goes in a huff! What a lot of fuss about nothing!
            (*He dances about and chants, beating time with his head
                as he turns about, on the front of the stage*)
There are too many babies, they're as common as flies,
All over the house with their hideous faces,
Making horrible noises and dirty wet places,
But the sooner you beat them the quicker they dies:
Hurrah for beating and tipping 'em over!
Your mother's gone crazy, your father's on clover....

Punch and Judy

(Re-enter JUDY *with a stick; she comes in behind and gives* PUNCH *a great wallop on the back of his head before he notices her)*

JUDY. I'll teach you to drop my child out of the window.

PUNCH. So-o-oftly, Judy, so-o-oftly! *(Rubbing the back of his head with his hands)* Don't be a fool, now. What are you at, now?

JUDY. What! You'll drop my poor baby out of the window again, will you? *(Hits him on the head)*

PUNCH. Oh, no, never again. *(She hits him again)* Softly I say, softly. Come now, Judy, a joke's a joke.

JUDY. Oh, you nasty cruel brute! *(Hits him again)* I'll teach you.

PUNCH. But I don't want to be taught. So it isn't a joke, eh? You're serious? You *mean* it do you?

JUDY. Yes *(hit)*, I *(hit)*, do *(hit)*.

PUNCH. I'm glad of that: I don't like practical jokes. *(She hits him)* I get your meaning now, you can leave off. *(she hits him)* So you won't leave off, eh?

JUDY. No, *(hit)*, I*(hit)*, won't *(hit)*.

PUNCH. All right then, I shall have to teach you.
(He snatches the stick and after a struggle he gets it and hits JUDY *on the head while she runs all over the stage to get out of his way)*
How do you like *that* lesson *(hit)* and *that* lesson *(hit)*, Judy my pretty dear?

JUDY. Oh pray, Mr Punch, no more!

PUNCH. Just one little lesson. *(hits her again)* One, two, three, and up to ten; that's the end of the first lesson, Amen. *(hit)*
*(*JUDY *falls down over the edge of the stage platform with one hand up to shield her head)*
Any more lessons?

JUDY. No, no! No more. *(Lifts up her head)*

PUNCH. *(knocks down her head)* I thought I'd soon quieten you down.

JUDY. *(lifts her head)* No!

PUNCH.   Yes!   (*Knocks her head down and follows up with blows until* JUDY *is dead*)
'At's a way to do it! Now, if you're satisfied, I am. (*Noticing that she doesn't move*) There, get up, Judy my dear; I won't hit you any more. None of your swinging the lead now, come along. (*Aside*) She's a playful little thing, full of fun! He! He! (*To* JUDY) You got a headache? Come on Judy, you're only asleep. Get up, I say! All right, then, get down. (*he tosses* JUDY's *body with his stick*) Oi de doi de doi! That's the way to do it! To lose a wife is to come into a fortune!

Who would ever put up with
a wife
If he had the chance to set
himself free
By doing her in with a rope
or a knife?

(*He throws away* JUDY's *body with his stick. Enter an* OLD BLIND MAN *while* PUNCH *is singing. He feels his way with his staff; he goes over the opposite side of the stage where he knocks*)

BLIND MAN.   Poor blind man, Mr Punch. I hope you'll bestow your charity. I hear you're very good and kind, to your wife, and to the poor, Mr Punch. Pray have pity upon me, and may you never know the loss of your tender eyes! (*He listens and hearing nobody coming, knocks again*) I lost my eyes in the war, Mr Punch. Pray, Mr Punch, have compassion upon the poor stone blind. (*He coughs and spits over the side*)

Only a halfpenny to buy something for my bad cough. I have a daughter to keep, Mr Punch. Only one halfpenny. (*Knocks again.* PUNCH *goes up to look at him closely and receives one of the knocks on his head*)

PUNCH. Hallo! You old blind blackguard, can't you see?

BLIND MAN. No, Mr Punch. Pray sir, give something to a poor blind man with a cough. (*coughs*)

PUNCH. Get along, you gurgling old villain. You'd only spend it on drink. I've nothing to spare for you. I'm only a poor widower.

BLIND MAN. Only a halfpenny. I have a daughter to keep. Oh dear, my cough is so bad. (*coughs in* PUNCH's *face*)

PUNCH. I'm blinded! You filthy old rascal! Was my face the dirtiest place you could find to spit in? Get away you nasty old blackguard. Get away! (PUNCH *seizes the* BLIND MAN's *staff and knocks him down inside the stage*) What a lot of uneducated people there are; what would they do if I didn't teach 'em? (*Sings, banging the side of the stage*)

If you don't want to grow up a fool
Come and be taught at Punch's school,
The lessons are read
Through the back of your head
And you get it all straight bending over a stool.
(*Enter* PRETTY POLLY *very gaily dressed*)

POLLY. Where is my father? My dear father.

PUNCH. (*aside*) O, what a beauty!

POLLY. Who killed my poor father? Oh! Oh! (*Cries*)

PUNCH. I can't tell you a lie, my dear. I did it with my little matchstick.

POLLY. Oh! Cruel wretch, why did you kill my father?

PUNCH. For your sake, my love.

POLLY. Oh, you barbarian!

Punch and Judy

PUNCH. Don't cry so, my dear. You will cry your pretty eyes out, and that would be a great pity.

POLLY. Oh! Oh! How could you kill him?

PUNCH. I asked him if I could marry you, and he said I couldn't. So I killed him. If you're going to go on taking on like that I shall have to cry too—Oh! oh! (*pretending to cry*) How sorry I am!

POLLY. Are you really sorry?

PUNCH. Yes, ever so sorry—look at the waterworks!

POLLY. (*aside*) What a handsome young man. It is a pity he should cry so. How the tears run down his beautiful long nose! Did you really kill my father out of love for me? (PUNCH *bows*) And are you sorry? PUNCH *holds his hands up under his chin and wags his shoulders coyly*) If you are sorry I must forgive you.

PUNCH. I could kill myself for love of you, *much more your father*.

POLLY. Do you then really love me?

PUNCH. Oi do! Oi do! Oi de diddley do!

POLLY. Then I must love you, too!

PUNCH. Wait a minute, I don't even know your name.

POLLY. (*turns her head aside and whispers inaudibly*)

PUNCH. What's that?

POLLY. (*whispers inaudibly again*)
PUNCH. Polly! Pretty Polly!

DUET

> When I think on you, my jewel,
> Don't wonder why my heart is sad:
> You're so lovely, yet so cruel,
> You're enough to drive me mad.
>
> See how much I love; take pity
> And relieve my bitter smart.
> Do you suppose you were made so pretty
> So you could break your lover's heart?

POLLY.　　　I love you so, I love you so,
　　　　　　I never will leave you, no, no, no.

PUNCH.　　　If I had all the wives of wise King Sol
　　　　　　I'd kill the lot off, and go off with Poll.
　　　　　　　　　(*Exeunt dancing*)

ACT II

Enter PUNCH

PUNCH. It's a very fine day.　　　(*Peeping out and looking at the sky*) I'll go fetch my horse and take a ride to visit my Pretty Poll.　　　(*Sings*)

(*Tune: 'Sally in our Alley'*)
> Of all the girls, that are so smart,
> There's none like Pretty Polly:
> She's the darling of my heart,
> For she's so plump and jolly.
> 　　　(*Enter the* GHOST *behind* PUNCH)

GHOST. So plump and jolly. (*disappears with horrible laughter*)
Ho ho ho!

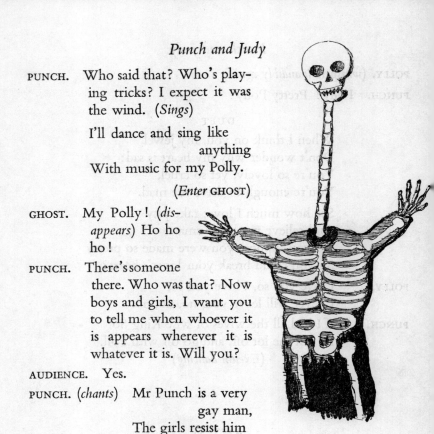

PUNCH. Who said that? Who's play-
ing tricks? I expect it was
the wind. (*Sings*)

I'll dance and sing like
                    anything
With music for my Polly.

(*Enter* GHOST)

GHOST. My Polly! (*dis-
appears*) Ho ho
ho !

PUNCH. There's someone
there. Who was that? Now
boys and girls, I want you
to tell me when whoever it
is appears wherever it is
whatever it is. Will you?

AUDIENCE. Yes.

PUNCH. (*chants*) Mr Punch is a very
                    gay man,
The girls resist him
    as much as they can
But they fall in the end for his nose and his
                                        chin.

(*Enter* GHOST. *And the audience shout 'Look out!' But by the
time* PUNCH *turns round the* GHOST *has gone*)

(PUNCH *goes on trying to sing his verse. Sometimes the* GHOST *is
behind one curtain, then behind another, but* PUNCH *can never find it.
He turns round suddenly to find the* GHOST *facing him. Its neck
suddenly grows longer, and twists round and round, and a grinning skull
looks down at him. Then he disappears*)

PUNCH. What's that?

AUDIENCE. The ghost !

PUNCH. I thought it was a big maggot jumping out of an apple.

(*Shouts after* GHOST) Who the devil are you, I should like to know, with your long neck? You may get it stretched for you, one of these days, you may be sure. That shan't stop me going after my dear Poll! (*He makes pugilistic squarings up with his fists: He goes out after* GHOST *and reappears with* HECTOR *his horse.* HECTOR *is very unruly*)

PUNCH. Whoa! ho! My lovely animal. Whoa! Hector. Stand still can't you and let me get me foot in the stirrup. Now off to see Pretty Poll!

(*While* PUNCH *is trying to mount, the horse runs away round the stage, and* PUNCH *sets off after him, catches him by the tail, and so stops him.* PUNCH *then mounts, by sitting on the front of the stage and with both his hands lifting one of his legs over the animal's back. At first,* HECTOR

Punch and Judy

goes pretty steadily, but soon he gallops. PUNCH *does not keep his seat very well and cries 'Whoa! Hector, Whoa!' but to no purpose, for the horse sets off at full gallop, jerking* PUNCH *at every stride with great violence.* PUNCH *lays hold round* HECTOR'S *neck, but suddenly the* GHOST *appears,* HECTOR *gives a loud whinny,* PUNCH *gives a loud shriek, and is thrown off on the platform.* HECTOR *disappears)*

PUNCH. Oh dear! Oh, lord! Help! Help! I'm murdered! I'm a dead man! Save my life! Doctor! Doctor! Doctor! Come and bring me to life again! I'm a dead man. Doctor!

(*Enter* DOCTOR)

DOCTOR. Who calls so loud?

PUNCH. Oh dear! Oh Lord! Murder!

DOCTOR. What is the matter? Good heavens, who is this? My good friend Mr Punch? Have you had an accident, Mr Punch? Or are you just taking a nap on the grass after dinner?

PUNCH. Oh Doctor! Doctor! I've been thrown off my horse! I've been killed.

DOCTOR. No, no, Mr Punch: not so bad as that, sir: you are not killed!

PUNCH. Well I can't speak, anyway.

DOCTOR. Where are you hurt? Is it here?   (*Flips* PUNCH *on the head, and* PUNCH *jerks his foot up*)

PUNCH. No, lower.

DOCTOR. Here?   (*Thumps him on the chest*)

PUNCH. No, further down.

DOCTOR. Here then?   (*Punches him in the back*)

PUNCH. No, lower still.

DOCTOR. Is your beautiful leg broken then? (*Cracks him on the shin*)

PUNCH. No, a little higher than me legs and a bit lower than me back.

DOCTOR. Well it must be, ah, let me see.... Ah. (*Bends down, but* PUNCH *kicks him in the eye*) Oh my eye! my eye! (*Exit*)

PUNCH. Ay, that's it—all my eye and Betty Martin.

> The Doctor is surely an ass, sirs
> To think I'm as brittle as glass, sirs,
> For I only fell down on the grass, sirs,
> And I've only a bone in my leg....

(*Enter* DOCTOR *behind with a stick and gives* PUNCH *several wallops on the back of his head*)

PUNCH. Hallo! Hallo! Doctor, what game are you up to now? What have you got there?

DOCTOR. I am a physician, Mr Punch. (*He makes a swing with the stick*) Physic is medicine, Mr Punch. (*Again*) People who have Betty Martin's disease must have physic. (*Hits him*)

PUNCH. Betty Martin's disease?

DOCTOR. You were swinging the lead, Mr Punch, and I'm swinging the physic. (*Hits him*)

PUNCH. I don't want any physic. Take it yourself.

DOCTOR. We never take our own physic if we can help it. (*Hits him*) A little more, Mr Punch, and you will soon be well. Physic! Physic! Physic! (*The* DOCTOR *knocks* PUNCH *all over the stage*)

PUNCH. Oh! Doctor! Doctor! No more! I'm ever so well now.

DOCTOR. Just one more little dose. (*Hits him*)

PUNCH. No more! Now you've treated me so well, do you know what I'm going to do?

DOCTOR. What Mr Punch? Leave me some money when you die?

PUNCH. No.

DOCTOR. Give me a lovely new motor-car?

PUNCH. No.

DOCTOR. What then?

PUNCH. Put that empty medicine bottle out on the step and I'll whisper.

DOCTOR. All right, Mr Punch. Now, no tricks now. Fair's fair.   (*Puts stick down on platform*)

PUNCH. No tricks. (*He grabs the stick*) And turn and turn about is fair too. I'm going to give you some physic. (*Hits him*)

DOCTOR. Hold off, Mr Punch. I don't want any physic, my good sir. I'm very well, thank you.

PUNCH. Oh no you're not.   (*Hits him*)

DOCTOR. Oh yes I am!

PUNCH. Oh no you're not. (*hit*) You're very ill. (*hit*) This will do you good. (*hit*) How do you like this physic? (*hit*) All free treatment. (*hit*) Physic! (*hit*) Physic!! (*hit*) Physic!!!   (*Hit*)

DOCTOR. (*gloomily*) Oh, Mr Punch, pray don't exceed the stated dose.

PUNCH. There's no dose for doctors. (*hit*) Doctors always die when they take their own medicine. (*hit*) Another little teaspoonful and you'll never want physic again. (*hit*) There, don't you feel the physic warming your inside? (*He pushes the stick into the* DOCTOR'S *stomach. The* DOCTOR *falls down dead, and* PUNCH *tosses his body away with the end of his stick*) Oi de doi de doi! That's the way to do it. (*He goes down to shout after the* DOCTOR) Now Doctor, you may cure yourself if you can.   (*Re-appears ringing a large sheep bell*)

Punch and Judy

Hurrah for old Punch, he's a medical fellow,
If you're feeling queer and you look a bit yellow
He'll knock down the quack
On the flat of his back
And treat you himself by beating you hollow!
With a physic for you and a physic for you,
He'll be ringing a bell as he
      sends you to hospi*tell*

(*Enter a* BLACK SERVANT *in a livery*)

SERVANT. Meester Punch, my master,
he say he no lika dat noise.

PUNCH. (*imitating the* SERVANT'S *Italian
accent*) Your master, he
say he no lika dat noise.
What noise?

SERVANT. Thees-a nasty noise.

PUNCH. D'you call that luverly
music a noise?

SERVANT. My master he no lika de
musica, Meester Punch, so
he'll have-a-da no more
noise near-a hees house.

PUNCH. Oh he don't, don't he. All
right-a. (*He runs round
the stage ringing his bell as loud as he can*)

SERVANT. You get-a away, I say, weeth thees-a nasty bell.

PUNCH. What bell?

SERVANT. That-a bell. (*Striking it with his hand*)

PUNCH. That's a good one. Do you call this a bell? (*Patting
it*) It's an organ.

SERVANT. And I say eet's a bell, a nasty bell.

PUNCH. I say it's an organ. (*Hits servant with it*) What do you
say it is now?

195

Punch and Judy

SERVANT.   An organ, Meester Punch.

PUNCH.   An organ? I say it's a fiddle. Can't you see it's a fiddle? (*Offers to strike him again*)

SERVANT.   Eet ees a feedle.

PUNCH. (*mocking him*)   'Eet ees a feedle.' I say it's a drum.

SERVANT.   Eeet ees a drum, Meester Punch.

PUNCH.   I say it's a trumpet.

SERVANT.   Well, so eet ees a trumpet-a. But bell, organ, feedle, drum or trumpet-a, my master, he say he no lika dat musica.

PUNCH.   Then bell, organ, fiddle, drum or trumpet, Meester Punch, he say your master is a fool.

SERVANT.   He say he no have-a-da musica near hees house.

PUNCH.   He's a fool I say not to like-a-da sweet musica. Go and tell him. Be off with you. (*hits him with the bell*) Get along. (*drives him round the stage, backwards; hitting him with the bell*) Be off!

(*Exit* SERVANT)

(PUNCH *continues to sing and dance and ring. Re-enter* SERVANT *with a stick.* PUNCH *sees him and goes behind the curtain at the side. The* SERVANT *hides the other side, but leaves the stick sticking out.* PUNCH *comes forward, puts down his bell very gently and creeps across the stage to see where the* SERVANT *is. Then he creeps back for his bell, takes it up, creeps over with it, and gives the* SERVANT *a great blow with it through the curtain. He goes out ringing his bell*)

SERVANT.   You dirty feelthy scoundrel, rascal, thief, vagabond blackgew-ard and ly-aire, you shall pay for thees!

(PUNCH *comes in and the* SERVANT *aims a blow at him but* PUNCH *goes out and the* SERVANT *misses.* PUNCH *comes back also armed with a stick which he hides behind his back. The* SERVANT *hits* PUNCH *on the head.* PUNCH *shakes his head*)

SERVANT.   Me teach-a you to reeng de nasty noisy bell near de gentil-men's houses!

196

PUNCH. (*produces his stick*)   Me teach-a you to lika da musica.
(*Hits* SERVANT. *Now they have a long fight, during which they change sticks and dodge round and round. But* PUNCH *wins and knocks* SERVANT *down on to the platform*)

SERVANT.   Oh dear! I am-a beaten, thees-a time!

PUNCH.   And I'm-a going to be a beating-a thees-a time! One! Two! three! There's some luverly music for you. This is my bell, (*hit*) this is my organ, (*hit*) this is my fiddle, (*hit*) this is my drum, (*hit*) and this is my trumpet. (*hit*) There! A whole band concert for you!

SERVANT.   No more! Me's-a feeneeshed.

PUNCH.   Quite feeneeshed?

SERVANT.   Yes, quite-a dead.

PUNCH.   Then there's the last for luck. Tom tiddleyom pom! Pom! Pom!   (*Hits him. Picks up the body by the legs, swings it round and throws it away*)

> Now I'll away to my Pretty Poll
> To show her my stick, with a folderol,
> That beats 'em and treats 'em
> For physic and music
> And turns everyone with his face to the wall!
>   (*Exit chanting*)

ACT III

(*Enter* PUNCH)

PUNCH.   Oh, I'm ever so happy. Or I would be, if only Polly would have me. But she won't. She says I'm too young and inexperienced. If she doesn't have me I shall get bad-tempered—me, the sweetest, gentlest fellow in the whole world.

Punch and Judy

When I think of you my jewel,
Don't wonder why my heart is sad...
You're so lovely, yet so cruel,
You're enough to drive me mad.
See how much I love, take pity....

(*Enter a* CONSTABLE)

CONSTABLE. Leave off singing, Mr Punch, for I've come to make you sing on the wrong side of your face.

PUNCH. Who the devil are you?

CONSTABLE. Don't you know me?

PUNCH. No, and I don't want to know you.

CONSTABLE. Oh, but you must. I'm the constable.

PUNCH. I didn't send for constable.

CONSTABLE. (*loftily*) I am sent for *you*.

PUNCH. I didn't want any constable. I can settle my own business without constable, thank you.

CONSTABLE. But the constable wants *you*.

PUNCH. The devil he does. What do you want me for?

CONSTABLE. You killed Mr Scaramouche. You knocked his head off his shoulders.

PUNCH. What's that got to do with you? If you stay here much longer, I'll do the same to you.

CONSTABLE. Now, now, Punch, don't talk to me like that. You have committed murder, and I have a warrant here for your arrest.

PUNCH. I don't want warrant. I'd like a rest, though.

CONSTABLE. I didn't say 'a rest', I said '*arrest*': jail, jug, ugh! (*He makes a noise and jerks his head and closes his eyes to mean that* PUNCH *will be hanged*)

PUNCH. What do you mean, 'arrest, jail, jug, ugh!'? Do that again, 'ugh!'.

CONSTABLE. 'Ugh!'

Punch and Judy

PUNCH.   You're not very well.

CONSTABLE.   Oh yes I am. I said 'ugh!', that's all.

PUNCH.   Oh no you're not. Look, you give me the warrant.

CONSTABLE.   Right.

PUNCH.   And I'll give you some physic. Then you can have a rest. Ready?

CONSTABLE.   Right.   (*Gives him warrant*)

PUNCH.   Do it again. Jail, jug, arrest....

CONSTABLE.   Ugh!

PUNCH.   Physic! (*hits him*) Physic! (*hit*) Physic. (*Knocks him down inside the booth, sings*) Jail, jug, warrant, arrest, take him away to have a long rest!

(*Enter an* OFFICER OF THE LAW *in a cocked hat with a cockade and a long pigtail*)

OFFICER.   Stop your noise, my fine fellow.

PUNCH.   Shan't.

OFFICER.   I'm an officer of the law.

PUNCH.   Did I say you weren't?

OFFICER.   You must go with me: you killed your wife and child.

PUNCH.   They were my own I suppose. I had a right to do what I liked with 'em.

OFFICER.   We'll see about that; I've come to take you up.

PUNCH.   And I'm going to take you down!   (PUNCH *knocks him down and sings as before*)

(*Enter* CHIEF JUSTICE OF ENGLAND *in a great wig*)

CHIEF JUSTICE.   Hello! Punch me boy!

PUNCH.   Hello! Who are you with a head like a cauliflower?

199

CHIEF JUSTICE. Don't you know me? I'm the Lord Chief Justice of England.

PUNCH. Lord Beef Pasties! I don't care if you're Lord Chancellor. You shan't have a chance to *chance the Law* on me. He! He!

CHIEF JUSTICE. But I shall have you in prison, Mr Punch. You're a murderer. You killed a poor old blind man. You've killed I don't know how many people.

PUNCH. If you don't know how many, you'd better go and find out.

CHIEF JUSTICE. That won't do, Punch me boy. You must come and be hanged.

PUNCH. I'll be hanged if I do. (*Knocks down the* CHIEF JUSTICE. PUNCH *dances and sings*)

O, Lord Beef Pasties, wearing
a wig,
With his cauliflower head a-
dancing a jig,
Will condemn you to death and
adjourn for his lunch,
But he shan't have the pleasure of sentencing Punch,
Hurrah for old Punch, he's a capital fellow....

(*Enter* OLD JACK KETCH, *the hangman, with a fur cap and black mask.* PUNCH *bumps into him as he dances, and steps back in alarm*)

PUNCH. My dear sir! I beg you a thousand pardons. Very sorry.

J. KETCH. Aye, you'll be sorry enough by the time I've done with you. Don't you know me?

PUNCH.   Oh, sir, I know you very well, and I hope you're very well, and Mrs Jack Ketch is very well.

J. KETCH.   Mr Punch, you're a very bad man. Why did you kill the Doctor?

PUNCH.   In self-defence.

J. KETCH.   That won't do.

PUNCH.   He tried to kill me.

J. KETCH.   How?

PUNCH.   With his damned physic. Shall I show you how, my dear sir?   (*Makes amusing little sallies with his stick*)

J. KETCH.   Come, come, Mr Punch. That's all gammon. You must come to prison. My name's Ketch, you know.

PUNCH.   Ketch that then.   (PUNCH *knocks down* JACK KETCH, *and sings*)

> Hangman, hangman,
> Climbing up so high,
> Let me take your ladder away
> And drop you on the sly.
> I'm sure you wouldn't like
> To waste that rope on me
> Wind it round your own windpipe
> And hang yourself on your own tree!
> With an oi de doi de diddley doi!

(*But enter behind, one after the other, the* CONSTABLE, *the* OFFICER, *and* JACK KETCH, *watched by the* LORD CHIEF JUSTICE. *The first three fall on* PUNCH, *and after a noisy struggle they pin him in a corner and carry him off, while he shouts 'Help! Murder!' etc.*)

(*The curtain at the back of the stage rises, and discovers* PUNCH *in prison, rubbing his nose against the bars and poking it through them*)

PUNCH.   Oh dear, what will become of poor old Punch now? My Pretty Poll, when shall I see you again?

Punch and Judy

Punch, when parted from his dear,
Sings all day a doleful tune:
I wish I had those rascals here
I'd settle their hash and be out of it soon.

(*Enter* JACK KETCH. *He fixes a gibbet on the platform of the stage and goes out*)

PUNCH. Well, look at that! That's very pretty. He must be a gardener. What a handsome tree he has planted, so you can see it from the window.

(*Enter the* CONSTABLE. *He places a ladder against the gibbet and goes out*)

PUNCH. Stop thief! Stop thief! He's a villain, he is. He'll come back and get up that ladder to steal the fruit off it, mark my words.

(*Enter* TWO MEN *with a coffin. They set it down on the platform and go out*)

PUNCH. What's that for, I wonder? Oh dear—I see now. What a fool not to see it. That's a large basket for the fruit to be put in.

(*Re-enter* JACK KETCH.)

J. KETCH. Now, Mr Punch, you may come out if you like.

PUNCH. Thank you very much, but I'm well off where I am. This is a very nice place, with a lovely view.

J. KETCH. Now, Mr Punch, I am come to make you suffer.

PUNCH. Make my supper! I don't want any supper.

J. KETCH. I did not say *supper*, Mr Punch. I said *suffer*. You must come out and be hanged.

PUNCH. You wouldn't be so cruel.

J. KETCH. Why were you so cruel as to commit so many murders?

PUNCH. But that's no reason why you should be cruel, too, and murder me.

J. KETCH. Come on, Punch.

PUNCH. I can't. I've got a bone in my leg, the Doctor said so.

Punch and Judy

J. KETCH.   You have a bone in your neck, Mr Punch, and that shall soon be broken.

(*He drags* PUNCH *out as* PUNCH *cries* 'Mercy! Mercy! I'll never do it again!')

J. KETCH.   Now, Mr Punch. No more delay. This is not a piece of firewood. It is not a tree. This is the gibbet, and this is the loop.

PUNCH.   Giblet soup! Giblet soup for supper. Oi-de-diddley-doi!

J. KETCH.   No, Mr Punch. *Gibbet* and *loop*. Now I want you to put your head in here.

PUNCH.   (*poking his head over the top of the gibbet*)   In here?

J. KETCH.   No, you silly man, in here.

PUNCH.   (*poking his head on one side of noose*)   In here?

J. KETCH.   No, you fool, in here.

PUNCH.   You mind who you're calling a fool. You see if you can do it yourself. Only show me how, and I'll do it directly.

J. KETCH.   Now look, stupid Mr Punch. You see my head? You see this loop? Now I put my head in here, so.   (*Puts his head in the loop*)

PUNCH.   And I pull the rope tight, so!   (*Pulls the rope so that* JACK KETCH *is drawn up to the top of the gibbet.* PUNCH *then swings on his body to make sure he is hanged*)
Oi de diddley doi. That's the way to do it! (*He then takes* JACK KETCH *down and puts his body in the coffin*)

(*Enter* TWO MEN *who remove the gibbet and place the coffin on it, dance with it on their shoulders grotesquely.* PUNCH *hides*)

MEN.   Punch is off the tree!
Punch is in the box!
To you, *from* me,
Put him to bed in his socks.

203

> Shove him in the ground,
> Shut him in the vault,
> Ups-a-daisy, slew him round
> Mind he doesn't bolt!
>
> Punch is off the tree!
> Punch is in the box!
> *To* you, *from* me,
> Put him to bed in his socks....
> *(Exeunt)*

PUNCH.   There they go. They think they've got Mr Punch safe and sound. But they've got old fur-cap Ketch. Oi-de-doi-de-doi! 'At's the way to do it. They're out! He's gone! I've done the trick. Jack Ketch is dead! I'm free. I don't care now if even Old Nick himself should come for me.

(He picks up his stick, and dances about beating time on the front of the stage with it, singing)

> Right fol de riddle lol,
> I'm the boy to do 'em all.
>   Here's a stick
>   To thump Old Nick,
> If he by chance should come to call.

(Enter the DEVIL. *He just peeps in at the corner of the stage and goes out)*

PUNCH.   *(very frightened and retreating as far as he can)*   Oh Lord! Talk of the Devil and he pops up his horns. There the old gentleman is, sure enough.

(A pause and dead silence, as PUNCH *continues to gaze at the spot where the* DEVIL *appeared. The* DEVIL *comes forward from the back)*

> Good, kind, Mr Devil. I never did you any harm, but all the bad in my power. There, don't come any nearer. How do you do, sir *(gathering courage)*. I hope you and all your respectable family are well? Much

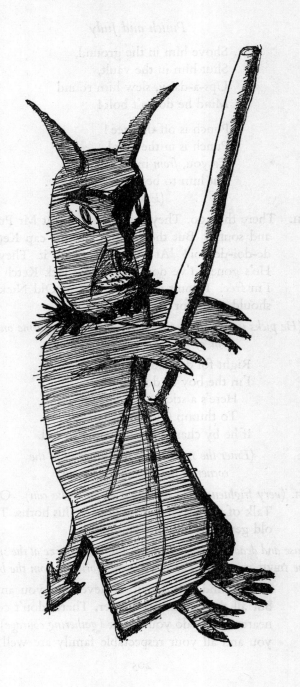

obliged for this visit. Good morning. I should be sorry to keep you, for I know you have a great deal of business when you're in London. (*The* DEVIL *advances*) Oh dear. What will become of me?

(*The* DEVIL *makes a dart at* PUNCH, *and* PUNCH *strikes at him, but the* DEVIL *avoids his blows, which only strike the boards. Exit* DEVIL)

PUNCH. Oi de doi de doi! He's off. He knows which side his bread is buttered; and toasted. He! he!

(*There is a terrible whirring noise.* PUNCH *cowers. The* DEVIL *comes in with a stick. They face one another. The* DEVIL *gives* PUNCH *a terrific whack on the head*)

PUNCH. Oh my head! What was that for? Pray Mr Devil, let's be friends. (*The* DEVIL *gives him another hit, whereupon* PUNCH *begins to take it in high dudgeon, and grows angry*) You must be a very stupid devil not to know your best friend when you see him. (DEVIL *hits him*) All right then! Let's see who is the best man—Punch or the Devil. (*They fight.* PUNCH *is badly knocked about at the beginning, but the* DEVIL *grows weary and* PUNCH *gets in some heavy blows. At last the* DEVIL *is knocked forward on to the platform and finished off with some great whacks.* PUNCH *then puts his stick up the* DEVIL's *tunic, hoists him on high and whirls him round in the air, crying*) Oi de doi de doi! That's the way to do it! The Devil's dead!

(*Curtain*)

THE END